PIXINSIGHT & STACKING WORKFLOWS FOR ASTROPHOTOGRAPHERS

Step-by-Step Preprocessing Guide to Transform Raw Frames into Deep-Sky Masterpieces

ANDREW HURST

DISCLAIMER

This book is an independent educational resource created to help astrophotographers learn image processing techniques. It is not affiliated with, endorsed by, or sponsored by Pleiades Astrophoto S.L. or the PixInsight development team. PixInsight® is a registered trademark of Pleiades Astrophoto S.L. All product names, trademarks, and registered trademarks mentioned in this book are the property of their respective owners and are used solely for identification and educational purposes.

While every effort has been made to ensure the accuracy of the information presented in this book, astrophotography workflows and software are constantly evolving. The author and publisher make no

representations or warranties with respect to the accuracy or completeness of the contents of this work and specifically disclaim all warranties, including without limitation warranties of fitness for a particular purpose. The advice and strategies contained herein may not be suitable for every situation.

CONTENT

INTRODUCTION

Why I Wrote This Book

I'll never forget the moment I first opened PixInsight. After weeks of hearing it called "the gold standard" for astrophotography processing, I had finally committed to the $280 license. I launched the software, imported my carefully collected light frames, darks, flats, and bias—and then stared at the screen, completely overwhelmed.

Where do I even start?

I clicked through menus. I opened processes with names like "ImageCalibration" and "WeightedBatchPreprocessing" that sounded promising but were filled with parameters I didn't understand. I watched a few tutorials, followed the steps exactly, and somehow ended up with calibrated frames that looked *worse* than my raw data. Stars were clipped. Backgrounds were uneven. My integrated master looked nothing like the beautiful examples I'd seen online.

After three frustrating evenings, I had produced exactly zero usable images.

Sound familiar?

If you've ever felt this way—excited by PixInsight's potential but intimidated by its complexity—this book is for you. You're not alone, and you're not doing anything wrong. PixInsight is genuinely difficult to learn, not because you're missing something obvious, but because preprocessing and stacking workflows involve dozens of interconnected decisions that aren't well explained anywhere.

The Preprocessing Problem

Here's what I discovered after months of trial, error, and digging through forum threads at 2 AM: most PixInsight books and tutorials skip past the hardest part.

They show you beautiful, stretched images with vibrant colors and sharp stars. They explain advanced processing techniques like deconvolution, narrowband palette combinations, and star reduction. But they gloss over the unglamorous foundation that makes all of that possible: **turning your raw frames into a clean, properly integrated master image**.

This preprocessing phase—calibration, stacking, and integration—is where most beginners struggle. It's where I struggled. And it's the part that matters most, because if you mess up preprocessing, no amount of clever post-processing will save your image.

The problem is that preprocessing is:

- **Technical**: It involves understanding bias frames, dark current, flat fields, rejection algorithms, and a dozen other concepts
- **Unforgiving**: Bad calibration frames can ruin your entire dataset
- **Poorly documented**: PixInsight's official documentation is comprehensive but overwhelming for beginners
- **Full of hidden pitfalls**: A single wrong setting (like FITS orientation) can silently destroy your results

I learned these lessons the hard way. I wasted perfectly good data because my dark frames were created at the wrong temperature. I spent hours troubleshooting why WBPP kept rejecting all my frames, only to discover my calibration masters were subtly wrong. I integrated beautiful data and got terrible results because I chose the wrong rejection algorithm.

Each mistake taught me something valuable—but it shouldn't have been so hard.

What This Book Covers (and What It Doesn't)

This book is laser-focused on one thing: **getting your raw astrophotography data through preprocessing and into a clean, integrated master image** that's ready for the creative processing phase.

You'll learn:

- How to create perfect master calibration frames (bias, darks, flats)
- How to calibrate your light frames without destroying them
- How to evaluate frame quality and select your best data
- How to register (align) your images properly
- How to stack (integrate) your frames using the right algorithms

- How to master WeightedBatchPreprocessing (WBPP) for automated workflows
- How to troubleshoot the most common preprocessing disasters
- How to handle complex scenarios: multiple sessions, meridian flips, mixed exposure times
- How to preprocess narrowband data properly

You won't find:

- Creative stretching and tone curve techniques (that's post-processing)
- Advanced topics like HDR composition or multi-scale processing
- Photoshop or alternative software workflows
- Planetary or lunar imaging techniques
- Deep dives into every PixInsight process (just preprocessing ones)

This focus is intentional. I want you to master the foundation before worrying about the fancy stuff. Once you can consistently produce clean, well-integrated masters, the rest of PixInsight becomes much less intimidating.

How This Book Is Different

There are other PixInsight books available. Some are comprehensive references that try to cover everything (and end up overwhelming). Others are too basic, giving you recipes to follow without explaining why. This book takes a different approach:

1. Practical Troubleshooting Focus
Every chapter includes real problems and their solutions. Not hypothetical scenarios—actual issues I encountered or saw others struggle with in forums. When something goes wrong (and it will), you'll know how to diagnose and fix it.

2. Decision Frameworks, Not Just Recipes
I don't just tell you to "set sigma low to 4 and sigma high to 2.5." I explain *why* you'd choose those values, when you'd change them, and what happens if you get it wrong. You'll develop judgment, not just follow steps.

3. Multiple Workflow Paths
Whether you prefer manual step-by-step control or automated WBPP workflows, I cover both approaches thoroughly. You'll understand when to use each and how to switch between them.

4. Camera-Specific Guidance
DSLRs, one-shot color cameras, and cooled monochrome cameras all require different approaches. I don't give generic advice that "sort of" works for everyone—I explain the specific considerations for different equipment.

5. Visual Learning
Screenshots, comparison images, before-and-after examples, flowcharts, and decision trees throughout. Seeing the problems and solutions makes everything clearer.

Who This Book Is For

You should read this book if you:

- Just purchased PixInsight and feel overwhelmed by where to start
- Have tried PixInsight tutorials but can't get consistent results
- Use WBPP but don't understand what it's actually doing
- Want to move from automatic workflows to more control
- Get frustrated when preprocessing fails and don't know why
- Need to combine data from multiple sessions or different equipment
- Want to understand the "why" behind every setting and parameter
- Are evaluating whether to invest in PixInsight

You might want a different book if you:

- Need comprehensive coverage of all PixInsight features
- Want creative post-processing and artistic techniques
- Are looking for planetary/lunar imaging workflows
- Prefer Photoshop or other software
- Already master preprocessing and need advanced topics

How to Use This Book

For Complete Beginners:
Start with Chapter 1 and read straight through. Each chapter builds on previous concepts. Follow along with your own data, even if you make mistakes—that's how you learn.

For WBPP Users Who Want Understanding:
Read Chapters 1-6 to understand what WBPP is doing automatically, then dive deep into Chapter 7. You'll use WBPP more effectively when you know what's happening under the hood.

For Troubleshooting Specific Problems:
Use the detailed table of contents and troubleshooting sections at the end of each chapter. The appendices include flowcharts for diagnosing common failures.

For Quick Reference:
Appendix A has quick-reference cards with optimal settings for common scenarios. Keep these handy during your processing sessions.

A Note on Software Versions

This book was written for PixInsight 1.8.9 and later, with WeightedBatchPreprocessing 2.5+. While the core concepts remain stable, PixInsight evolves continuously. Check the book's website ([Your Website]) for updates, corrections, and notes on newer versions.

My Promise to You

I can't promise that PixInsight will suddenly become easy. The software is complex because astrophotography itself is complex. But I can promise that after working through this book, you'll:

- Understand *why* preprocessing steps matter and what each one accomplishes
- Know how to create quality calibration frames that improve (not harm) your data
- Be able to diagnose and fix common preprocessing problems
- Have confidence in your workflow, whether manual or automated
- Spend less time frustrated and more time enjoying the hobby
- Produce consistently better integrated masters as your foundation for processing

Most importantly, you'll stop feeling like PixInsight is a mysterious black box. You'll understand what's happening to your data at every step, which gives you the power to make informed decisions and troubleshoot problems independently.

Let's Begin

Astrophotography is one of the most rewarding hobbies I've ever pursued. There's something magical about capturing photons that traveled across the cosmos and turning them into a beautiful image. But that magic requires technical skill, and technical skill requires good teaching.

Consider me your guide through the preprocessing wilderness. I've made the mistakes so you don't have to. I've spent countless hours debugging problems so I can help you avoid them. And I've organized everything I've learned into a clear path from confusion to competence.

Your clear skies and steady tracking await. Let's turn those raw frames into something beautiful.

Clear skies and good stacking,

Andrew Hurst

CHAPTER 1: UNDERSTANDING PIXINSIGHT PREPROCESSING

The Foundation for Deep-Sky Mastery

You've just finished an incredible imaging session. Three hours of perfectly tracked exposures under dark skies. Your target was centered, your focus was sharp, and the seeing was excellent. You pack up your equipment with excitement, imagining the stunning image you're about to create.

The next day, you sit down at your computer, import your frames into PixInsight, and... now what?

You have 150 light frames, 50 dark frames, 50 flat frames, and 100 bias frames staring back at you. PixInsight's interface is open, showing dozens of processes with cryptic names. You know these raw frames contain an amazing image—you just need to extract it.

This is the moment where many astrophotographers feel overwhelmed. Welcome to preprocessing.

1.1 Why Preprocessing Matters More Than You Think

Let me share a hard truth I learned after ruining several datasets: **your final image quality is determined more by preprocessing than by creative processing**.

You can spend hours perfecting your color balance, adjusting curves, and applying sophisticated deconvolution—but if your preprocessing was flawed, you're polishing a fundamentally damaged image. It's like trying to sharpen a photograph that was out of focus when you took it. No amount of post-processing can recover data that was destroyed during calibration and stacking.

The Hidden Quality Killers

Preprocessing is where several critical quality factors are either preserved or destroyed:

Signal-to-Noise Ratio (SNR)
Every frame you capture contains both signal (light from your target) and noise (random variations from the sensor, electronics, and sky background). Proper stacking mathematically reduces noise while preserving signal. The result: fainter details become visible, colors become more accurate, and your image gains depth.

Poor stacking, however, can actually degrade your SNR. I've seen images where satellite trails weren't properly rejected, leaving bright streaks. I've seen cases where too-aggressive rejection algorithms threw away good data along with the bad. Most commonly, I've seen improper calibration add noise instead of removing it.

Dynamic Range
Your sensor captures a specific range of brightness levels, from the faintest photons to the brightest stars. Preprocessing should preserve this entire range, allowing you to stretch the data later without clipping shadows or blowing out highlights.

Bad calibration—particularly with dark frames—can crush your shadows, pushing faint nebulosity into complete blackness. Once that data is lost, it's gone forever. I learned this lesson when I discovered my dark frames were taken at a different temperature than my lights, causing massive oversubtraction that destroyed faint details in my target.

Star Profiles
Sharp, round stars make images look professional. Bloated, elongated, or misshapen stars scream "amateur" (and not in a good way). Most star quality issues originate in preprocessing: poor registration causing slight misalignment, bad flat fields creating vignetting artifacts, or incorrect stacking parameters that blur fine details.

Color Accuracy
For one-shot color cameras, improper debayering or calibration can introduce color casts that are difficult or impossible to correct later. I once spent an hour trying to fix magenta stars in post-processing, only to realize the problem originated from calibrating my flats incorrectly.

Artifact Removal
Hot pixels, cold pixels, amp glow, gradients, satellite trails, cosmic rays—your raw frames are filled with artifacts that have nothing to do with your target. Preprocessing is your opportunity to remove these systematically. Miss this step, and you'll spend hours cloning them out manually in post-processing (trust me, I've been there).

How Bad Preprocessing Ruins Perfect Sessions

Let me tell you about the night I captured the Horsehead Nebula. Conditions were perfect: Bortle 2 skies, no moon, excellent seeing. I collected four hours of Ha data with my new cooled monochrome camera. The histogram during capture looked amazing—I knew I had something special.

Then I processed it.

I used my usual dark frames—the ones I'd shot the previous month. The calibration seemed fine at first glance, but when I integrated the frames, something was wrong. The background was mottled with a walking noise pattern. Faint areas of nebulosity that should have been visible simply weren't there. Stars had dark rings around them.

After two days of troubleshooting, I discovered the problem: my dark frames were taken when the camera was 10°C warmer than during the imaging session. Dark current is temperature-dependent, so my "calibration" was actually adding noise instead of removing it. I had to reshoot the entire target.

Four hours of imaging. Four months waiting for good weather and moonless skies. All wasted because I didn't understand a fundamental preprocessing principle: dark frames must match the temperature of your light frames.

That expensive lesson taught me something valuable: **preprocessing isn't a formality you rush through to get to the "real" processing. Preprocessing IS the real processing**. Everything else is just presentation.

What Professional Astrophotographers Know

Professional observatories and research institutions spend enormous resources on calibration and data reduction. The Hubble Space Telescope, for example, has entire teams dedicated to preprocessing. Why? Because they understand that scientific data must be properly calibrated to be meaningful.

Amateur astrophotographers sometimes think we're different—that we're making "art" not "science," so we can be less rigorous. This is a mistake. The physics of sensors, noise, and signal processing doesn't change based on whether you're doing science or art. The techniques that produce scientifically accurate data also produce aesthetically beautiful images.

The best amateur astrophotographers I know spend 60-70% of their processing time on preprocessing. They carefully craft their master calibration frames. They review every light frame for quality. They experiment with stacking parameters to find the optimal settings for their data. Only after the preprocessing is perfect do they move to creative stretching and color work.

This investment pays off. Their images have a quality that's immediately recognizable: clean backgrounds, sharp stars, excellent color, and faint details that others miss. The secret isn't expensive equipment or exotic filters—it's mastery of preprocessing.

1.2 The PixInsight Preprocessing Pipeline Explained

Now that you understand why preprocessing matters, let's look at what actually happens to your data during the preprocessing pipeline. Think of this as a factory assembly line, where raw frames enter one end and a clean, integrated master image emerges from the other.

The Complete Journey: Raw to Master

Here's the big picture of what we're going to accomplish:

Stage 1: Master Calibration Frame Creation

- Raw bias frames → Stacked master bias
- Raw dark frames → Stacked master dark
- Raw flat frames → Calibrated flats → Stacked master flat

Stage 2: Light Frame Calibration

- Raw light frames + master bias/dark/flat → Calibrated light frames
- Calibrated lights → Cosmetically corrected lights (hot/cold pixel removal)

Stage 3: Quality Assessment

- Calibrated lights → Quality metrics (FWHM, eccentricity, SNR)
- Select best frames, identify reference frame

Stage 4: Registration

- Calibrated lights + reference frame → Aligned/registered lights
- Generate drizzle data (optional)

Stage 5: Integration

- Registered lights → Integrated master image
- Apply weighting, normalization, rejection algorithms

Stage 6: Drizzle (Optional)

- Registered lights + drizzle data → Super-resolution master

The output of this pipeline is your master light frame—a single, high-quality image that represents all the signal from all your individual exposures, with noise reduced and artifacts removed. This master is still in linear (unstretched) form, ready for creative processing.

Understanding Linear vs. Non-Linear Workflows

This is one of the most important concepts in astrophotography processing, and it confuses many beginners: **preprocessing must happen on linear data**.

What is Linear Data?

When your camera sensor captures light, each pixel records the number of photons that hit it during the exposure. This relationship is linear: twice as many photons = twice the pixel value. Your raw frames are linear data—they directly represent the physical light that entered your telescope.

Linear images look terrible when you open them. They appear almost completely black with a few bright stars. This is because most of the signal from deep-sky objects is very faint, occupying only the darkest 10-20% of your sensor's range. Human vision is logarithmic (we perceive relative brightness differences, not absolute ones), so linear data doesn't display well on screens.

What is Non-Linear Data?

Non-linear data has been "stretched"—pixel values have been remapped using curves or histogram transformations to make the image displayable. Faint nebulosity becomes visible, mid-tones occupy the middle of the tonal range, and stars remain bright but not clipped.

Stretching is essential for viewing and artistic processing, but it fundamentally changes the mathematical relationships in your data. Once stretched, you can no longer perform certain mathematical operations correctly.

Why Preprocessing Requires Linear Data

Consider what happens when you calibrate with a dark frame. You're literally subtracting one set of pixel values from another:

Calibrated Pixel = Light Pixel - Dark Pixel

This simple subtraction works correctly only when the pixel values have a linear relationship to the actual signal. If your data is non-linear (stretched), this subtraction becomes mathematically meaningless.

The same principle applies to all preprocessing operations:

- **Flat field correction** divides your light frames by the flat frame
- **Image integration** averages or median-combines pixel values
- **Rejection algorithms** compare pixel values statistically
- **Normalization** scales images to match background levels

All of these operations assume linear relationships. Perform them on non-linear data, and you'll get incorrect, unpredictable results.

The Golden Rule

Here's the rule that will keep you out of trouble: **All preprocessing happens on linear data. Only after you have a final integrated master do you stretch to non-linear.**

I mark this boundary clearly in my workflow. Everything before stretching is "preprocessing" or "calibration." Everything after stretching is "post-processing" or "creative processing." Never mix the two.

The Data Flow Map

Let me give you a visual mental model for how your data flows through PixInsight. Imagine three distinct zones:

Zone 1: Raw Data (Linear)

- Your camera's raw files
- FITS or XISF files from your acquisition software
- Completely unprocessed sensor data
- Looks almost black when viewed

Zone 2: Calibrated Linear Data

- Master calibration frames
- Calibrated light frames
- Registered light frames
- Still linear, still looks very dark
- Ready for mathematical operations

Zone 3: Integrated Master (Linear)

- Single master light frame
- All frames combined
- Noise reduced
- Still linear—this is critical
- The end product of preprocessing

The Boundary: Stretching to Non-Linear

After you have your integrated master, you stretch it. This is the point of no return—once stretched, you cannot go back to valid linear data. All future processing is creative/artistic, not mathematical/calibration.

Zone 4: Stretched Data (Non-Linear)

- Visible nebulosity and details
- Suitable for creative processing
- Color correction, contrast, sharpening
- The data you'll eventually export as a JPEG or TIFF

Understanding this flow prevents countless mistakes. Every time you're about to perform an operation in PixInsight, ask yourself: "Is my data still linear?" If yes, you can do calibration and preprocessing operations. If no (it's been stretched), you're in the creative processing phase.

1.3 Manual vs. Automated Workflows: Making the Right Choice

PixInsight offers two fundamentally different approaches to preprocessing: manual step-by-step processing and automated batch processing using the WeightedBatchPreprocessing (WBPP) script. Understanding when to use each approach will save you enormous amounts of time and frustration.

The Manual Workflow: Step-by-Step Control

The manual workflow means you execute each preprocessing step individually:

1. Open ImageIntegration, load your bias frames, set parameters, create master bias
2. Open ImageIntegration again, load dark frames, create master dark
3. Open ImageCalibration, load flat frames, calibrate them with master bias/dark
4. Open ImageIntegration again, create master flat
5. Open ImageCalibration, load light frames, calibrate with all masters
6. Open SubframeSelector, measure frame quality
7. Open StarAlignment, register your frames
8. Open ImageIntegration, stack your registered frames

Each step requires understanding what you're doing, selecting appropriate parameters, and verifying the results before moving to the next step.

Advantages of Manual Workflow:

Complete Control
You make every decision explicitly. You can see exactly what each step does to your data. When something goes wrong, you know exactly which step caused the problem.

Learning Opportunity
By doing each step manually, you learn what's actually happening. You understand parameters, not just recipes. This knowledge becomes invaluable for troubleshooting.

Flexibility for Complex Data
Some datasets don't fit standard patterns. Maybe you

have multiple exposure times, or you're combining data from different sessions, or your calibration frames need special handling. Manual workflow gives you the flexibility to handle these cases.

Easier Troubleshooting
When something goes wrong (and it will), you can isolate the problem to a specific step. You can redo just that step with different parameters without starting over.

Disadvantages of Manual Workflow:

Time Consuming
The manual workflow takes longer, especially once you know what you're doing. What WBPP can do in one click requires you to open and configure 8-10 different processes.

More Opportunities for Error
Each step you perform manually is an opportunity to make a mistake: selecting the wrong files, forgetting a parameter, using the wrong reference frame.

Requires More Knowledge
You need to understand what each process does and how to configure it. The learning curve is steeper.

Easy to Lose Track
With dozens of intermediate files (calibrated lights, registered lights, etc.), file management becomes challenging. You can easily get confused about which files are at which stage.

The WBPP Workflow: One-Click Automation

WeightedBatchPreprocessing (WBPP) is PixInsight's automation powerhouse. You load all your raw files—bias, darks, flats, and lights—configure some parameters, click Run, and WBPP executes the entire preprocessing pipeline automatically.

WBPP handles:

- Creating all master calibration frames
- Calibrating your light frames
- Optionally performing cosmetic correction
- Measuring frame quality and calculating weights
- Registering frames
- Integrating them into a final master
- Optionally performing drizzle integration

Advantages of WBPP:

Speed and Efficiency
Once configured, WBPP can process an entire session in one pass. What takes an hour manually takes 5 minutes of setup time (though WBPP itself may run for hours on large datasets).

Consistency
WBPP applies the same logic and parameters to all your data. You won't accidentally forget a step or use different settings for different frames.

Quality Weighting
WBPP automatically measures frame quality (like SubframeSelector) and uses this information to weight frames during integration. Better frames contribute more to the final master.

Fewer Files to Manage
WBPP creates intermediate files in organized folders, and you can often delete them after processing. You don't have to track dozens of intermediate images.

Disadvantages of WBPP:

Black Box Problem
When WBPP works, it's magical. When it fails, troubleshooting is difficult because you don't know exactly which internal step failed or why.

Less Flexibility
WBPP makes certain assumptions about your data and workflow. If your situation doesn't fit its model (unusual calibration needs, complex multi-session data), you may need to work around its limitations or use manual processing.

Harder to Learn From
Because WBPP does everything automatically, you don't learn what each step does. If you only ever use WBPP, you won't develop the deep understanding needed for troubleshooting or advanced techniques.

All or Nothing
If one part of your workflow needs special handling, you can't easily do "most of it with WBPP and one step manually." You generally need to choose one approach or the other (though hybrid approaches exist).

So Which Should You Use?

Here's my honest recommendation, based on my own learning journey and watching others:

Start with Manual Workflow

If you're new to PixInsight, force yourself to do manual preprocessing for your first 5-10 targets.

Yes, it's slower. Yes, it's more tedious. But you'll learn:

- What master calibration frames should look like
- How to verify calibration worked correctly
- Why registration fails and how to fix it
- How different rejection algorithms affect your results

This knowledge is foundational. Once you have it, WBPP becomes a tool you understand rather than magic you trust blindly.

Transition to WBPP for Standard Workflows

Once you understand preprocessing, use WBPP for straightforward, standard sessions:

- Single night, single target
- Standard camera (DSLR, OSC, or mono)
- Normal calibration frames
- Single filter (or filter sets that process independently)

WBPP will save you time while producing excellent results, and you'll understand what it's doing under the hood.

Return to Manual for Complex Scenarios

Use manual processing when:

- Combining data from multiple sessions/instruments
- Mixing different exposure times for HDR
- Your calibration frames need special handling
- WBPP keeps failing or producing unexpected results
- You need precise control over specific parameters

The Hybrid Approach

Many experienced users (including me) use a hybrid approach:

- Use WBPP to create master calibration frames quickly
- Use WBPP for initial calibration of lights
- Switch to manual for registration and integration where I want precise control

Or:

- Use manual processing to get everything to the registered frames stage
- Use WBPP or ImageIntegration for the final stacking

The flexibility to move between approaches comes from understanding both. That's why learning the manual workflow first is so valuable.

The WBPP Paradox

Here's something interesting I've noticed: the users who get the most value from WBPP are the ones who don't need it. They understand preprocessing so well that they can configure WBPP properly and diagnose problems when they occur.

Users who jump straight to WBPP without understanding the underlying steps struggle when things go wrong. They post in forums saying "WBPP rejected all my frames, help!" but can't provide details about their calibration quality or settings because they never learned to evaluate these things.

Don't let WBPP become a crutch. Learn the fundamentals, then use automation to speed up what you already understand.

1.4 Essential PixInsight Setup and Configuration

Before we dive into actual preprocessing, let's make sure PixInsight is configured properly. A few critical settings can make the difference between smooth workflows and constant frustration.

First Launch: Initial Configuration

When you first launch PixInsight, the interface is empty—just a gray workspace. This is intentional; PixInsight doesn't assume how you want to work. Let's set up a productive environment.

Setting Format Preferences

One of the most critical (and commonly missed) settings is FITS orientation. This setting must be consistent across all your preprocessing, or you'll get wrong calibration results.

Go to: **Edit > Global Preferences > Image Format > FITS**

Find the parameter: **Vertical Orientation**

Set it to: **Bottom-up (traditional)**

Why? This ensures PixInsight interprets FITS files the same way your acquisition software probably created them. Mismatched orientation means your flat fields apply to the wrong part of the image, your darks don't subtract correctly, and calibration fails in subtle but destructive ways.

I wasted an entire dataset before discovering this setting. My flats seemed fine visually, but after calibration, I had vignetting in places where there shouldn't be any. The problem? My acquisition software created bottom-up FITS files, but PixInsight was reading them as top-down. The flat was being applied upside-down.

IMPORTANT: Once you set this, never change it unless you're deliberately reprocessing old data with different orientation. Consistency is critical.

Setting DSLR Raw Preferences (If Using DSLR)

If you shoot with a DSLR, configure how PixInsight handles your raw files:

Go to: **Edit > Global Preferences > Image Format > RAW**

Key settings:

- **Interpolation Algorithm**: VNG or AHD (high quality debayering)
- **Create Super Pixels**: Unchecked (we want full resolution)
- **Output Images**: Raw Bayer CFA (we'll debayer manually during preprocessing)

Wait—why "Raw Bayer CFA" instead of automatically debayered RGB?

Because debayering is a preprocessing step, and we want control over when it happens. If PixInsight automatically debayers your raw files when opening them, you've already made an irreversible decision. By keeping them as Bayer CFA, we can debayer at the optimal point in our workflow (usually after calibration).

Memory and Performance Settings

Go to: **Edit > Global Preferences > Process > Memory**

- **Maximum Concurrent Threads**: Set to your CPU's physical core count (not logical cores/hyperthreads)
- **Enable Scratch Files**: Checked (allows PixInsight to use disk space when RAM fills up)

These settings help PixInsight use your hardware efficiently. More threads = faster processing, but only up to your physical core count. Beyond that, you're just adding overhead.

Workspace Organization

PixInsight's workspace is infinitely customizable, which is powerful but initially confusing. Let me share an organizational approach that works well for preprocessing.

Creating a Default Workspace

Think of your workspace as your workbench. You want your most-used tools readily accessible.

Essential processes for preprocessing (add these to your workspace):

- **ImageIntegration** (you'll use this constantly)
- **ImageCalibration** (ditto)
- **SubframeSelector** (for quality assessment)
- **StarAlignment** (for registration)
- **DynamicCrop** (for cropping borders after registration)

How to add them permanently:

1. Open the process: **Process > [Category] > [Process Name]**
2. Click the triangle icon at the bottom of the process window
3. Select "Add to Workspace"

Now these processes appear as icons in your workspace. Click an icon to open a new instance of that process with your default parameters.

Process Icons: Your Saved Configurations

Here's a powerful feature: you can save process configurations as icons and drag them onto images to apply them instantly.

For example, after you figure out the perfect ImageIntegration settings for your DSLR images, you can save those settings as an icon. Next time, you just drag that icon onto your image stack, and all your parameters are already set.

To create a process icon:

1. Configure a process with your desired settings
2. Click the triangle at the bottom
3. Select "New Instance"
4. A small icon appears at the bottom of your workspace

Drag this icon onto an image to apply those exact settings. Or drag it to the workspace to save it permanently.

I have dozens of saved process instances for different scenarios: "DSLR Integration," "Mono Integration - Ha," "Dark Frame Stack," etc. This saves enormous amounts of time once you develop your standard workflows.

File Management Strategy

PixInsight doesn't force a file organization system on you, which means you need to create your own. Here's what works well:

/2025-01-15_M42_Session/

/Raw/

/Lights/

/Darks/

/Flats/

/Bias/

/Masters/

master_bias.xisf

master_dark_300s.xisf

master_flat_L.xisf

/Calibrated/

[calibrated light frames]

/Registered/

[registered frames]

/Integrated/

master_light_300s.xisf

/Final/

[stretched and processed images]

This structure keeps everything organized and makes it clear which stage each file represents. When WBPP runs, I point it to output folders within this structure.

Understanding File Formats: FITS vs. XISF vs. RAW

PixInsight works with several file formats, and choosing the right one matters for both quality and efficiency.

FITS (Flexible Image Transport System)

- **Pros**: Universal astronomy format, preserves all metadata, works with every astronomy program
- **Cons**: Large files, slower to read/write, limited to 32-bit
- **When to use**: Sharing data, archiving, compatibility with other software

XISF (Extensible Image Serialization Format)

- **Pros**: PixInsight's native format, fast, efficient, 64-bit capable, excellent compression
- **Cons**: Only PixInsight and a few other programs can read it
- **When to use**: Working within PixInsight, intermediate files you won't share

Camera RAW (CR2, NEF, ARW, etc.)

- **Pros**: Direct from camera, maximum original data
- **Cons**: Large, slow to process, requires conversion
- **When to use**: Initial import, then often converted to FITS or XISF

My Workflow Strategy:

For DSLR shooting, I keep originals as RAW, but I immediately convert them to XISF when importing to PixInsight. Why? Memory efficiency.

PixInsight must load entire RAW files into RAM at once. But XISF files can be loaded progressively in chunks. When you're integrating 150 RAW files, you might run out of RAM. Convert to XISF first, and suddenly the same operation succeeds.

To batch convert:

1. **Process > Format Conversion > FormatConversion**
2. Add your RAW files
3. Set output format to XISF
4. Enable compression
5. Run

For cooled cameras that output FITS, I often keep them as FITS through preprocessing because that's their native format. But for final masters that I'll work with repeatedly, I save to XISF for speed.

Memory Management Fundamentals

PixInsight's biggest limitation is often RAM. Understanding how to manage memory prevents countless headaches.

How Much RAM Do You Need?

The quick formula: You need about 10× the size of your raw image files to process them comfortably.

If your camera produces 50MB raw files and you're stacking 100 of them:

- Conservative: 50MB × 100 × 10 = 50GB RAM (safe, unlikely to run out)
- Minimum: 50MB × 100 × 3 = 15GB RAM (tight, may struggle)

I have 32GB RAM and can comfortably process typical DSLR datasets (24MP sensor). But when I upgraded to a 26MP cooled camera, I started hitting memory limits during large integrations.

Memory-Saving Strategies:

Use XISF instead of RAW
As mentioned, XISF can be loaded progressively, reducing peak memory usage.

Process in batches
Instead of integrating 200 frames at once, integrate 50 at a time, then integrate the four resulting masters. The math is slightly less optimal, but it's better than crashing.

Use 32-bit instead of 64-bit
PixInsight can process images in 32-bit or 64-bit per channel. During preprocessing, 32-bit is usually sufficient and uses half the memory. You can always convert to 64-bit later for post-processing if needed.

To force 32-bit: In file open dialogs, check "Import as 32-bit"

Enable scratch files
We enabled this in preferences earlier. If PixInsight runs out of RAM, it will swap data to disk. This is slow but prevents crashing.

Close unused images
Every open image consumes RAM. After you've created your master bias and saved it, close the individual bias frames. Keep your workspace tidy.

Monitor Memory Usage

At the bottom right of PixInsight's interface, you'll see memory usage stats. Watch this during large operations. If you see it climbing toward your physical RAM limit, you know you need to adjust your approach.

1.5 Understanding Your Data Before You Process

The final step before diving into actual preprocessing is understanding what you're working with. Different cameras, filters, and acquisition approaches require different preprocessing strategies.

Camera Types and Their Implications

DSLR and Mirrorless Cameras

Characteristics:

- Bayer CFA (Color Filter Array) sensor
- Fixed gain/ISO settings
- Potential for amp glow
- Limited cooling (ambient temperature only)
- Often shoot RAW proprietary formats

Preprocessing implications:

- Must handle debayering (converting Bayer CFA to RGB)
- ISO settings must match between lights and darks

- Dark frames highly temperature-dependent
- Amp glow often visible in corners (long exposures)
- Built-in dark subtraction must be disabled

One-Shot Color (OSC) Dedicated Astro Cameras

Characteristics:

- Bayer CFA sensor (like DSLR)
- Gain instead of ISO (continuous, not steps)
- Active cooling (TEC)
- No built-in processing
- Outputs FITS files directly

Preprocessing implications:

- Must handle debayering
- Gain must match between lights and darks (usually fixed for all subs)
- Temperature-controlled darks (much more consistent than DSLR)
- Less amp glow (better electronics)
- May have different gain settings for different targets

Cooled Monochrome Cameras

Characteristics:

- No color filter array (true monochrome)
- Active cooling
- Lower read noise than OSC
- Requires filters for color imaging
- Ultimate sensitivity

Preprocessing implications:

- No debayering needed (much simpler)
- Each filter requires separate master flats
- Can mix integration times from different filters more easily
- Temperature control critical for darks
- Calibration must be done per-filter

Understanding which type of camera you have determines your entire preprocessing approach. The workflows are similar but not identical, and details matter.

Filter Considerations

Broadband Filters

These pass most of the visible spectrum (RGB, luminance, or clear filters):

- Create "natural color" images
- Affected by light pollution
- Faster (more light gets through)
- Simpler processing

Narrowband Filters

These pass only specific wavelengths (Ha, OIII, SII):

- Reject light pollution effectively
- Much slower (less light passes)
- Enable imaging from urban areas
- Require longer exposures
- Need special color combination techniques
- Each filter processed completely separately

Dual-Band and Tri-Band Filters

These pass multiple narrowband lines in one filter (Ha+OIII, for example):

- Compromise between narrowband selectivity and speed
- Only work with OSC cameras
- Special processing to separate the channels
- More complex preprocessing

Your filter determines:

- How many separate calibration/integration workflows you need
- What kind of color combination you'll do later
- How you'll handle flats (one per filter)

Reading FITS Headers

FITS files contain metadata that tells you critical information about how the image was captured. Learning to read this metadata prevents preprocessing mistakes.

To view FITS headers in PixInsight:

1. Open an image
2. **Process > FITS Header**
3. A window displays all the metadata

Key fields to check:

EXPOSURE or **EXPTIME**

- Exposure length in seconds
- Must match between lights and darks

GAIN or **ISOSPEED**

- Camera gain setting
- Must match between lights and darks for proper calibration

SET-TEMP or **CCD-TEMP**

- Target or actual sensor temperature
- Critical for cooled cameras
- Should be consistent across all lights in a session

FILTER

- Which filter was in place
- Essential for organizing multi-filter datasets

DATE-OBS

- When the exposure was taken
- Useful for sorting and organizing

IMAGETYP

- Light, Dark, Flat, or Bias
- Used by WBPP for automatic file classification

Why check headers? Because sometimes acquisition software mislabels files, or you forget which darks match which lights. The headers don't lie—they tell you exactly what each file represents.

I once spent an hour trying to figure out why my calibration looked terrible. Eventually, I checked the FITS headers and discovered that my "180-second darks" were actually 300-second darks that got renamed incorrectly. The header caught the mistake immediately.

Pre-Flight Checklist

Before starting any preprocessing session, verify:

- [ ] All raw files are present (lights, darks, flats, bias)
- [ ] File counts make sense (enough calibration frames)
- [ ] FITS orientation preference is set correctly
- [ ] You understand your camera type and have the right workflow planned
- [ ] FITS headers confirm exposure times, gains/ISOs, and filters match where they should
- [ ] You have enough disk space (preprocessing can generate 3-5× your raw data size temporarily)
- [ ] You have enough RAM for your dataset size (or have a batch processing plan)
- [ ] Your workspace is organized with essential processes accessible
- [ ] You've backed up your raw data (preprocessing is non-destructive, but better safe than sorry)

With this foundation in place, you're ready to create your first master calibration frames.

Key Takeaways from Chapter 1:

- Preprocessing determines final image quality more than creative processing
- All preprocessing must happen on linear data; stretching comes later
- Manual workflow teaches fundamentals; WBPP provides speed once you understand the basics
- Critical settings (FITS orientation) must be configured before processing anything

- Understanding your camera type and filters determines your workflow approach
- FITS headers contain the truth about your data—check them when in doubt

In Chapter 2, we'll dive deep into creating perfect master calibration frames—the foundation that everything else builds upon.

CHAPTER 2: MASTER CALIBRATION FRAMES

Building the Foundation of Quality

Imagine building a house on a foundation with cracks and uneven settling. No matter how beautiful the construction above, the house will have problems. Master calibration frames are your preprocessing foundation. Get them right, and everything else becomes easier. Get them wrong, and you'll fight quality issues through every subsequent step.

I learned this lesson the expensive way—with ruined datasets and countless hours of troubleshooting. Let me save you that pain by showing you exactly how to create calibration frames that improve your images instead of damaging them.

2.1 The Science Behind Calibration

Before we dive into the "how," let's understand the "why." What are we actually calibrating, and what does each type of calibration frame do?

The Problem: Your Sensor Isn't Perfect

When photons from a distant nebula strike your camera sensor, you'd love to believe that each pixel accurately records the light it received. Unfortunately, reality is messier:

Problem 1: Bias and Read Noise

Even with zero exposure time and the shutter closed, your sensor reads out small voltage values. This is the "bias level"—the electronic baseline that the camera adds to every pixel before amplification. It varies slightly across the sensor and between readouts.

Think of it like the hum of an amplifier before you plug in a guitar. It's always there, it's not signal, and you need to account for it.

Problem 2: Dark Current

Sensors generate electrons thermally, not from photons. The warmer the sensor, the more thermal electrons accumulate during an exposure. This "dark current" varies across the sensor (some pixels are hotter than others) and increases with exposure time and temperature.

At room temperature with long exposures, dark current can be significant—sometimes overwhelming the actual signal from faint targets. This is why cooled cameras exist.

Problem 3: Pixel Response Variations (Vignetting and Dust)

Not every pixel sees the same amount of light, even if the incoming light is perfectly uniform:

- **Vignetting**: Optical systems naturally darken toward the edges
- **Dust**: Specks on filters, sensors, or optical surfaces block light
- **Pixel sensitivity**: Manufacturing variations mean some pixels are more sensitive than others

If you photograph a perfectly uniform blank field, your sensor won't record it as uniform. It'll show darkening at the corners, donuts from dust specks, and subtle variations from pixel-to-pixel differences.

The Solution: Mathematical Correction

Calibration frames allow us to measure these problems and correct for them mathematically:

- **Bias frames** measure the electronic baseline with zero exposure
- **Dark frames** measure bias plus dark current at a specific exposure time and temperature
- **Flat frames** measure the relative response of each pixel to uniform illumination

By capturing these separately, we can subtract or divide them from our light frames to remove the systematic errors they represent.

The Mathematical Truth

Here's the fundamental equation of image calibration:

Calibrated Light = (Light - Dark) / Flat

Or more precisely:

Calibrated Light = (Light - Dark) / (Flat - Dark)

In words: We subtract the dark signal from both the light frame and flat frame, then divide the corrected light by the corrected flat. This removes both additive errors (bias and dark current) and multiplicative errors (vignetting and dust).

This math only works if:

1. Your dark frames accurately represent the dark current in your lights
2. Your flat frames accurately represent the pixel response variations
3. Everything is done on linear data

Get any of these wrong, and the calibration degrades your image instead of improving it.

What Each Calibration Type Removes

Let me be very specific about what each type does:

Bias Frames

- **Remove**: Electronic bias offset, read noise pattern
- **Characteristics**: Zero exposure time, minimal signal
- **Temperature**: Doesn't matter (not thermal)
- **Stacking**: Average of many frames to reduce read noise

Dark Frames

- **Remove**: Bias + dark current + hot pixels
- **Characteristics**: Same exposure as lights, sensor covered
- **Temperature**: Must match lights (dark current is temperature-dependent)
- **Stacking**: Average of many frames with rejection to remove cosmic rays

Flat Frames

- **Remove**: Vignetting, dust shadows, pixel response variations
- **Characteristics**: Uniformly illuminated field at correct focus
- **Exposure**: Typically very short (1-5 seconds)
- **Stacking**: Average of many frames to reduce noise

Understanding these distinctions helps you troubleshoot when calibration goes wrong. If you see hot pixels after calibration, your darks weren't adequate. If you see vignetting, your flats failed. If everything looks too dark, you might have over-subtracted with darks.

2.2 Creating Master Bias Frames

Bias frames are the simplest calibration frames to shoot, but there are still details that matter.

Why Bias Frames Matter

Bias frames serve two purposes:

1. **Dark calibration**: Used to calibrate your dark frames before stacking them
2. **Light calibration**: Can be used to calibrate lights when you're not using dark frames (rare in deep-sky work)

For cooled cameras with excellent dark performance, you might use "bias + dark optimized" rather than just darks. For DSLRs, you'll almost always use "darks only" which already include bias.

Some astrophotographers skip bias frames entirely and just use darks. This works, but using bias frames properly gives you slightly better results by reducing the noise contribution from your calibration.

How Many Bias Frames Do You Need?

The answer depends on your camera's read noise, but here's a practical guide:

Minimum: 50 frames
This is the bare minimum for decent noise reduction in your master bias. Below this, you're not getting much benefit from stacking.

Good: 100 frames
This is my standard. At 100 frames, you've reduced the read noise by $\sqrt{100} = 10\times$. The master bias is smooth and low-noise.

Overkill: 200+ frames
Beyond 100-150 frames, you're seeing diminishing returns. Your master bias noise becomes negligible compared to other error sources. Unless you have a very high read-noise camera or you're doing absolute precision work, this is probably excessive.

My recommendation: Shoot 100 bias frames and keep them as a library for your camera. Bias frames don't change unless your camera settings change, so one good set can last months or years.

Shooting Bias Frames Correctly

The process is simple, but don't skip steps:

Step 1: Camera Configuration

- Set the same gain/ISO you'll use for light frames
- Set the shortest exposure time your camera allows (usually 1/4000s or 1/8000s for DSLRs)
- Cover the lens/telescope (cap, cloth, doesn't matter—no light should enter)
- Disable any in-camera dark frame subtraction or noise reduction

Step 2: Take Your Exposures

- Shoot 100 frames (takes 5-10 minutes)
- Keep your camera at a stable temperature if possible (doesn't matter as much as for darks)
- Save as RAW (for DSLRs) or FITS (for dedicated astro cameras)

Step 3: Verify

- Open a few random bias frames in PixInsight
- They should look almost uniformly gray with just noise
- Histogram should be narrow and centered around a low value
- If you see patterns, gradients, or unusual structure, something went wrong

Common mistakes:

Light leaked in: You'll see gradients or bright areas

- **Wrong exposure time**: Too long (>1 second) and you're getting dark current
- **Camera processing enabled**: In-camera processing will corrupt the bias

Creating Your Master Bias in PixInsight

Now let's stack those 100 bias frames into a single, low-noise master.

Step 1: Open ImageIntegration

Process > ImageIntegration > ImageIntegration

You'll see a complex interface with many tabs and parameters. Don't worry—for bias frames, most settings are straightforward.

Step 2: Add Your Bias Frames

Click the "Add Files" button (document icon with green plus) and navigate to your bias frames. Select all 100 and click Open.

If your frames are DSLR RAW files, PixInsight will load them as Bayer CFA images (you'll see a green-tinted monochrome image). This is correct—don't debayer bias frames.

Step 3: Configure Integration Parameters

Image Integration Tab:

- Leave all files selected (checkmarks)
- No need to set weights (bias frames should be equal quality)

Format Hints:

- If using DSLR RAW: Set "Input hints" to your camera's pattern (usually RGGB)
- This tells PixInsight how to interpret the Bayer pattern

Integration Tab:

- **Combination**: Average
  - Why average? Because we want to reduce noise. Median is more robust against outliers, but for bias frames, outliers are rare.
- **Normalization**: No normalization
  - Why? Bias frames should already have identical levels. Normalization would be meaningless.
- **Weights**: Don't care
  - We're averaging all frames equally
- **Generate integrated image**: Checked
  - We want the actual master bias output
- **Evaluate noise**: Checked
  - Useful to see if our master bias has low noise
- **Generate rejection maps**: Unchecked (optional)
  - Not critical for bias frames

Rejection Tab:

- **Rejection algorithm**: Percentile clipping
    - Why? With 100 frames, percentile clipping handles the occasional cosmic ray or hot pixel read
- **Percentile low**: 0.2
- **Percentile high**: 0.1
    - These reject the bottom 20% and top 10% at each pixel, protecting against outliers

Step 4: Execute the Integration

Click the blue checkmark (Execute) button at the bottom left. ImageIntegration will process your frames. Depending on your computer and file sizes, this might take 30 seconds to a few minutes.

When complete, a new image window opens: your master bias.

Step 5: Evaluate Your Master Bias

Your master bias should:

- Look uniformly gray with subtle, smooth variations
- Have very low noise (smooth, not grainy)
- Show no obvious patterns, gradients, or artifacts
- Have a narrow histogram at the low end

To check statistics:

Process > Image > Statistics

Click the green arrow (Apply to Active Window).

Key values to check:

- **Mean**: Should be low, typically 500-1000 for 16-bit images
- **Standard deviation**: Should be small (the noise level)
- **Median**: Should be close to mean (symmetric distribution)

If your sigma (standard deviation) is very low (< 50 for 16-bit), you have an excellent master bias. If it's high (> 200), you might need more frames or have a high read-noise camera.

Step 6: Save Your Master Bias

File > Save As

- **Format**: XISF (PixInsight's native format) or FITS
- **Filename**: Something descriptive like master_bias_gain100_100frames.xisf
- Include the gain/ISO in the filename so you know what it's for

Important naming convention: I include three pieces of information in calibration master filenames:

1. Type (bias, dark, flat)
2. Camera settings (gain/ISO, exposure time if applicable)
3. Frame count (optional but useful)

Example: master_bias_ISO1600_100frames.xisf

This makes it easy to select the correct master later.

Creating a Superbias (Advanced)

A "superbias" is a special master bias that's even lower in noise. The concept: instead of using a regular master bias, you create a synthetic bias by taking the median of many master biases.

Why bother?

If you're doing extremely precise photometry or pushing very faint data, the noise in your master bias becomes a limiting factor. A superbias can reduce this noise by another factor of 2-3×.

How to create one:

1. Create 5-10 separate master biases (each from 50-100 frames)
2. Open ImageIntegration again
3. Add these master biases as input files
4. Combination: Median (not average)
5. No rejection needed
6. Generate the output

The result is a superbias with exceptionally low noise.

Do you need this?

Honestly, probably not for typical deep-sky imaging. The difference between a good master bias (100 frames) and a superbias is marginal compared to other error sources. I mention it for completeness, but don't feel obligated to do this.

I use regular master biases for everything, and my images are limited by other factors long before bias noise becomes an issue.

Troubleshooting Bias Frame Problems

Problem: Master bias shows horizontal or vertical lines

This is "banding" or "pattern noise" from your camera's electronics. Some cameras (especially DSLRs at low ISO) have significant banding.

Solution:

- Shoot more frames (200+) to reduce the banding through averaging
- Consider using a higher ISO/gain where banding is less prominent
- For severe banding, you might need to use specialized banding reduction tools

Problem: Master bias has bright spots or hot pixels

Bias frames shouldn't have hot pixels—those require exposure time to accumulate. If you see them, either:

- Your "bias" frames had exposure time (check your camera settings)
- You have stuck/defective pixels that are always bright

Solution:

- Verify your exposure time was truly minimal
- Use cosmetic correction or a defect map to handle stuck pixels

Problem: Master bias looks "choppy" or very noisy

You didn't stack enough frames, or your camera has very high read noise.

Solution:

- Stack more frames (150-200)
- If your camera just has high read noise, accept it—your darks and lights will overwhelm this noise anyway

Problem: Different bias frames have very different histograms

Your camera or acquisition software might be applying some auto-processing or your camera settings weren't consistent.

Solution:

- Ensure in-camera processing is completely disabled
- Verify settings were identical for all bias frames
- Reshoot if necessary

2.3 Creating Master Dark Frames

Dark frames are more complex than bias frames because they're temperature and exposure-time dependent. Get these wrong, and calibration actively harms your images.

Dark Frame Theory: Exposure Time and Temperature

Dark current accumulates over time and increases exponentially with temperature. This creates two critical rules:

Rule 1: Dark frames must match exposure time

If your lights are 300-second exposures, your darks must be 300 seconds. A 180-second dark won't have the right amount of dark current for a 300-second light. Using mismatched exposure times causes over-subtraction or under-subtraction.

Exception: Dark scaling

PixInsight can "optimize" or scale dark frames to match different exposure times, assuming dark current is linear with time. This works reasonably well for moderate differences (180s dark for 240s light), but not for large differences (60s dark for 600s light).

My recommendation: Always shoot darks that match your light exposure times exactly. Dark scaling is a last resort, not a standard practice.

Rule 2: Dark frames must match temperature

Dark current roughly doubles for every 6-8°C increase in temperature. This means:

- A dark shot at 20°C has twice the dark current of one at 13°C

- A dark shot at 0°C has 8× less dark current than one at 20°C

For DSLR cameras (no active cooling), this is a challenge. Your darks must be shot at approximately the same ambient temperature as your lights. If you image on a 15°C night and your darks were shot on a 25°C afternoon, the calibration will fail.

For cooled cameras with temperature control, this problem disappears. Set your camera to -10°C, and your darks shot at -10°C will always match your lights shot at -10°C, regardless of ambient conditions.

How Many Dark Frames Are Enough?

Dark frames are noisier than bias frames (they have all the read noise plus dark current noise plus cosmic rays), so you need more of them.

Minimum: 30 **frames**
This is acceptable for well-controlled cameras (cooled astro cameras). It's barely adequate for DSLRs.

Good: 50 **frames**
My standard for cooled cameras. Provides good noise reduction and cosmic ray rejection.

Better: 100+ **frames**
For DSLRs or whenever you want maximum quality. The extra frames significantly improve rejection of transient artifacts.

The longer your exposure time, the more dark frames you need. A 10-minute dark exposure will have multiple cosmic ray hits per frame, so you need many frames to reject these outliers effectively.

Shooting Dark Frames Correctly

Timing: Ideally, shoot darks during or immediately after your light frames, when temperature conditions match. If shooting on a different day, match ambient/sensor temperature as closely as possible.

Camera Configuration:

- **Same gain/ISO as lights**: Critical
- **Same exposure time as lights**: Critical
- **Same temperature as lights**: Critical (for cooled cameras, set the target temp)
- **Cover telescope/lens**: No light should enter
- **Disable in-camera processing**: Especially dark frame subtraction

For DSLRs:

- Shoot darks at night when temperatures match your imaging sessions
- If temperature changes during your light session, you might need darks from multiple temperature ranges
- Some people maintain a "dark library" with darks shot at different temperatures (10°C, 15°C, 20°C, etc.) and select the closest match

For Cooled Cameras:

- Set the target temperature (typically -10°C to -20°C)
- Wait for temperature to stabilize before starting
- You can shoot darks during daytime (since it's light-sealed and temp-controlled)
- Build a library of darks at your standard temperature for each exposure time you use

Cosmic Ray Warning:

Long exposure darks (5+ minutes) will contain cosmic ray hits—bright spots or trails from high-energy particles striking the sensor. These are random and won't be in the same place on different frames. This is why we need rejection when stacking darks.

Never use a single dark frame—you'd be applying cosmic ray artifacts to all your lights!

Creating Your Master Dark in PixInsight

The process is similar to master bias, with a few key differences.

Step 1: Open ImageIntegration

Process > ImageIntegration > ImageIntegration

Step 2: Add Your Dark Frames

Click "Add Files" and select all your dark frames (50-100 of them).

Step 3: Configure Integration Parameters

Integration Tab:

- **Combination**: Average
    - Average is standard for darks
- **Normalization**: No normalization

- Dark frames should already have matching levels

- **Weights**: Don't care
- **Evaluate noise**: Checked

Rejection Tab:

Here's where darks differ from bias. We need strong rejection to remove cosmic rays and hot pixel outliers.

- **Rejection algorithm**: Sigma clipping
  - Why? We have enough frames (50+) for iterative sigma clipping to work well
- **Sigma low**: 4.0
- **Sigma high**: 3.0
  - These are fairly tight, aggressively rejecting outliers
- **Clip low pixels**: Checked
- **Clip high pixels**: Checked
  - We want to reject both unusually dark and unusually bright pixels (cosmic rays are bright outliers)

For 100+ frames, you could use Linear fit clipping, which is even more robust:

- **Rejection algorithm**: Linear fit clipping
- Uses a more sophisticated statistical model
- Excellent for large frame counts

Step 4: Execute Integration

Click Execute and wait. Darks take longer to process than bias frames due to the rejection algorithm iterations.

Step 5: Evaluate Your Master Dark

Your master dark should:

- Look similar to individual darks but smoother (less noise)
- Show the characteristic "hot pixel" pattern of your sensor
- Have higher signal than master bias (dark current added)
- Show no obvious cosmic ray hits (rejection worked)

Check rejection maps (optional):

If you enabled rejection maps, PixInsight creates two additional images showing which pixels were rejected as outliers. These should show scattered bright pixels (cosmic rays) fairly uniformly distributed. If you see large areas rejected, something went wrong.

Check statistics:

Process > Image > Statistics

The median and mean should be higher than your master bias (dark current signal added). Standard deviation will also be higher (dark current adds noise).

Step 6: Save Your Master Dark

File > Save As

Filename example: master_dark_300s_ISO1600_-10C_50frames.xisf

Include:

- Exposure time (critical!)
- Gain/ISO
- Temperature (especially for DSLR library darks)
- Frame count (optional)

Dark Optimization: Scaling Darks

PixInsight has an "Optimize" option in ImageCalibration that scales darks to match exposure times. Should you use it?

When optimization works:

- Your dark and light exposure times differ by less than 2×
- Your camera has linear dark current behavior
- You shot darks at the correct temperature

When optimization fails:

- Large exposure time differences (2× or more)
- Non-linear dark current (some cameras)
- Temperature mismatch (optimization can't fix this)

My recommendation: Avoid relying on optimization. Shoot darks that match your lights exactly. Use optimization only as a fallback when you discover you don't have matching darks.

Dark Libraries: Building a Reference Set

Once you have a cooled camera with temperature control, you can build a reusable dark library:

Concept: Shoot master darks once for each combination of:

- Exposure time (60s, 180s, 300s, 600s, etc.)
- Gain setting (if you use multiple gains)
- Temperature (your standard imaging temp, like -10°C)

Benefits:

- Never shoot darks again (just reuse your library)
- Darks are built from hundreds of frames (maximum quality)
- Saves time during imaging sessions

Limitations:

- Only works with temperature-controlled cameras
- Assumes your camera's dark characteristics don't change over time (they do, very slowly)
- Should remake library annually or if you notice degradation

Structure example:

/Dark_Library/

master_dark_60s_gain100_-10C_200frames.xisf

master_dark_180s_gain100_-10C_200frames.xisf

master_dark_300s_gain100_-10C_200frames.xisf

master_dark_600s_gain100_-10C_150frames.xisf

For DSLRs, dark libraries are less practical because you can't control temperature precisely. You'd need libraries at multiple temperatures, which becomes unwieldy.

Troubleshooting Dark Frame Problems

Problem: Master dark has streaks or trails

These are cosmic rays that weren't rejected. You need:

- More dark frames (100+)
- Tighter rejection parameters (lower sigma values)
- Better rejection algorithm (try linear fit clipping)

Problem: Master dark looks almost identical to master bias

Either:

- Your exposure time was too short (not enough dark current accumulated)
- Your camera is very cold (minimal dark current)
- Something went wrong with your dark exposures

Check a few individual darks. If they also look like bias frames, reshoot with verified exposure time and covered optics.

Problem: Master dark has strange patterns or gradients

Possible causes:

- Light leaked in during dark exposure (cover wasn't effective)
- Amp glow (visible in camera corners, especially DSLRs)
- Camera sensor artifacts

Amp glow is normal for some cameras—it's an electronic artifact that appears as brightening in corners. Your master dark should capture this so it gets subtracted from lights.

Problem: After dark calibration, my light frames look too dark or have negative pixels

Over-subtraction! Your darks had too much signal. Causes:

- Temperature mismatch (darks shot warmer than lights)
- Exposure time mismatch
- Dark optimization applied incorrectly

Solution:

- Verify dark and light metadata match (check FITS headers)

- Reshoot darks at correct temperature/exposure
- Disable optimization if you were using it inappropriately

2.4 Creating Master Flat Frames

Flat frames are the most misunderstood and often poorly executed calibration frames. They're also the most immediately visible in your final image—bad flats create obvious vignetting or dust donuts.

Flat Field Theory: Why Flats Are Critical

Imagine photographing a perfectly uniform white surface. You'd expect every pixel to record the same value. But that's not what happens:

- **Vignetting**: Your optical system naturally darkens toward the edges
- **Dust shadows**: Any dust on your sensor, filters, or optical surfaces creates dark spots
- **Pixel response variations**: Some pixels are more or less sensitive than others
- **Filter effects**: Filters can introduce additional vignetting or shadows

Flat frames capture all these variations by photographing an evenly illuminated field. When you divide your light frame by the (normalized) flat, you remove these multiplicative errors.

The flat field equation:

Corrected = Light / Flat

Or more precisely (since flats also contain bias and dark):

Corrected = (Light - Dark) / (Flat - Dark)

The Flat Frame Challenge

Flats are harder than bias or darks because you need an evenly illuminated surface at the correct focus. "Evenly illuminated" is tricker than it sounds—any gradient or pattern in your light source becomes imprinted on your flats and transferred to your calibrated images.

Common flat shooting methods:

- **T-shirt over telescope** (dawn/dusk sky)
- **Light panel** (electroluminescent or LED)
- **White surface with flashlight** (tricky to get even)
- **Twilight sky** (no obstructions)

Each method has trade-offs. The key is achieving even illumination without gradients.

Shooting Flat Frames: The Critical Details

Timing and Focus:

- **Same focus as light frames**: Critical! Dust shadows change with focus distance
- **Same optical configuration**: Same filters, rotators, focal reducers, etc.
- **Timing**: Either during your imaging session (keep everything the same) or shortly before/after

Exposure Settings:

- **Exposure time**: Adjust to place histogram peak around 40-50% (mid-tones)
  - Why? Too dark and you're noise-limited; too bright and you risk saturation
  - Flats are usually 1-5 seconds depending on your light source brightness
- **Gain/ISO**: Same as your light frames
  - Some people use lower ISO for flats (less noise), but matching is safest

Illumination Requirements:

- **Even**: No gradients, hotspots, or patterns
- **Bright enough**: Should require only 1-5 second exposures for proper exposure
- **Color neutral**: Especially important for OSC/DSLR cameras
- **Diffuse**: Not directional (avoid point sources that create patterns)

How Many Flats?

- **Minimum**: 20 frames
- **Good**: 30 frames
- **Better**: 50 frames

Flats are relatively quick to shoot (short exposures), so there's little reason not to shoot 30-50. More frames = better noise reduction.

Flat Shooting Methods: Pros and Cons

Method 1: T-Shirt/White Cloth + Dawn/Dusk Sky

Process: Stretch a white t-shirt over your telescope aperture, point at blank dawn or dusk sky, shoot flats.

Pros:

- Free (just need a white shirt)
- Sky provides very even illumination
- Easy to get correct exposure

Cons:

- Requires dawn/dusk timing (inconvenient)
- Weather-dependent
- Color temperature changes as sky brightens/darkens
- Can't easily reshoot if you forgot

Method 2: Electroluminescent (EL) Panel

Process: Place EL panel over telescope aperture, shoot flats.

Pros:

- Very even illumination
- Controllable brightness
- Can shoot anytime
- Repeatable

Cons:

- Requires purchasing EL panel ($30-60)
- Color temperature may not be neutral
- Some panels show subtle patterns

Method 3: LED Light Panel

Process: Use photography LED panel or DIY lightbox, point telescope at it, shoot flats.

Pros:

- Very bright (fast flats)
- Adjustable brightness
- Can shoot anytime
- Repeatable

Cons:

- Can create hotspots or gradients if not diffused properly
- Some LED panels have flicker (use constant-on mode)
- Requires building or buying light source

Method 4: Twilight Sky (No Cover)

Process: Point telescope at blank sky during twilight, shoot flats.

Pros:

- Perfectly even illumination
- Natural color temperature
- No equipment needed

Cons:

- Timing critical (short window)
- Must avoid stars, clouds, moon
- Can't change filters easily (lose timing window)

My Recommendation:

Start with the t-shirt method—it works well and costs nothing. As you get more serious, invest in a quality EL panel for convenience and repeatability.

For astrophotographers with multiple filters (LRGB or narrowband), a lightbox is almost essential. Shooting flats for 5-7 filters during twilight is nearly impossible; you need a controlled light source.

Creating Your Master Flat in PixInsight

Flats require one extra step: they must be calibrated with your master bias and/or dark before stacking.

Why calibrate flats?

Your flat frames contain:

- The flat field information we want
- Bias offset (we don't want)
- Dark current (we don't want)

If we stack flats without calibrating them, we're averaging together the bias and dark current noise, which then gets applied to all our light frames. By calibrating flats first, we remove these components.

Step 1: Calibrate Your Flat Frames

Process > ImageCalibration > ImageCalibration

Configuration:

- **Target Frames**: Add your flat frames (30-50 of them)
- **Enable**: Master Bias (check, then select your master bias file)
- **Enable**: Master Dark (only if using darks for flats)
    - Many people skip darks for flats since flat exposures are very short (1-5 seconds, minimal dark current)
    - If your flats are longer than 10 seconds, consider using a matching dark
- **Master Flat**: Leave disabled (we're calibrating flats, not applying them)

Calibration Settings:

- **Calibrate only**: Unchecked (we want calibrated output files)
- **Pedestal**: 0 (default)
- **Optimization**: Disabled for bias; consider enabling for darks if exposures differ

Output:

- Set an output directory for calibrated flats
- Use a suffix like "_cal" so you know they're calibrated

Click Execute. ImageCalibration processes each flat frame, subtracting bias (and dark if enabled).

Step 2: Integrate Calibrated Flats

Now open ImageIntegration:

Process > ImageIntegration > ImageIntegration

Add Files: Your calibrated flat frames (from the output directory of step 1)

Integration Parameters:

- **Combination**: Average
    - Standard for flats
- **Normalization**: Multiplicative
    - Critical for flats! This normalizes each flat to the same scale before averaging
    - Why? Flat exposures might vary slightly (1.5s, 1.8s, 1.4s) due to changing light conditions
    - Multiplicative normalization scales them all to match before combining
- **Weights**: Don't care
- **Rejection**: Sigma clipping (3.0, 2.5)
    - Flats shouldn't have many outliers, but gentle rejection handles the occasional dust mote or transient artifact

Execute Integration

A new image appears: your master flat.

Step 3: Evaluate Your Master Flat

Your master flat should:

- Show obvious vignetting (darkening toward corners)
- Reveal any dust shadows (dark donuts, usually round or slightly elongated)
- Be smooth and free of gradients or patterns
- Have high SNR (smooth, not grainy)

Checking the histogram:

The histogram should have a strong peak somewhere in the mid-tones (not clipped on either end). If your flats were properly exposed, the master flat histogram should look good.

Visual inspection:

Use **Process > All > ScreenTransferFunction** (STF) to stretch the master flat for viewing. You should clearly see vignetting. Any dust shadows should be obvious as well-defined dark circles.

If you see:

- **Gradients**: Your illumination wasn't even; reshoot flats with better light source
- **Patterns or rings**: Optical artifacts or light source problems
- **Extreme noise**: You didn't shoot enough flats or they were too dim

Step 4: Save Your Master Flat

File > Save As

Filename example: master_flat_L_filter_50frames.xisf

Include:

- Filter designation (critical if you have multiple filters!)
- Frame count (optional)

CRITICAL NOTE: You need separate master flats for every filter you use. A luminance flat won't work for Ha. An OIII flat won't work for SII. The dust shadows are at different focus distances for different wavelengths.

If you shoot LRGB, you need four master flats. If you shoot narrowband SHO, you need three master flats. Plan accordingly.

Flat Frame Best Practices

Practice 1: Shoot flats every session

Even with permanent setups, dust accumulates and focus changes slightly. Fresh flats ensure accurate calibration.

Practice 2: Verify flat quality immediately

Check your flats while still at the telescope. Open one in PixInsight and apply STF. Can you see vignetting? Is illumination even? If not, reshoot before packing up.

Practice 3: Take more flats than you think you need

Flats are quick. Shoot 50 instead of 30. The extra time is negligible, and more frames = better master.

Practice 4: Document your flat shooting method

Take notes on:

- Light source used
- Exposure time required
- Any issues or observations
- Weather/ambient conditions

This helps you replicate success and troubleshoot failures.

Practice 5: Don't reuse flats across sessions with different equipment configurations

If you change filters, rotate your camera, adjust focus, or modify anything optical, you need new flats. Flats from last month with different focus will create artifacts.

Troubleshooting Flat Frame Problems

Problem: Master flat shows obvious gradients

Your illumination source wasn't even. Solutions:

- If using t-shirt: ensure it's stretched tight and evenly lit
- If using panel: add more diffusion between panel and telescope
- If using sky: shoot at different time when sky is more uniform
- Consider switching to a better light source (EL panel)

Problem: No vignetting visible in master flat

Either:

- Your optics genuinely have no vignetting (rare but possible with well-designed systems)
- Your flat illumination was so uneven that it masked vignetting
- Something went very wrong

Check individual flat frames. They should show vignetting if your system has any (most do).

Problem: Dust donuts look blurry or wrong size

Focus wasn't the same as during light frame capture. Dust shadows change appearance with focus distance. Solutions:

- Reshoot flats at exact focus position used for lights
- Use parfocal rings or marks to ensure repeatable focus
- Focus on a star before shooting flats

Problem: After flat calibration, my images have bright or dark rings

This is one of the most common flat problems. Causes:

- Flats shot at different focus than lights (dust shadows misaligned)
- Flats shot with different filter than lights
- Flat exposure was too dim or too bright (poor SNR)
- Flat illumination had patterns that got imprinted

Solutions:

- Verify flats match lights in all optical configuration
- Reshoot flats with better exposure/illumination
- Check individual calibrated flats to see if problem is in the flat or elsewhere

Problem: Master flat has strange bright or dark spots

Possible causes:

- Dust on flat illumination source moved during shooting
- Sensor contamination that changed during flat shooting
- Gradient from uneven illumination

Examine individual flats. If the spots move between frames, your light source or sensor had moving dust. If they're consistent, you have a real feature (dust on sensor/filter) that your flats will correctly remove.

2.5 Advanced Calibration Considerations

Now that you understand the basics, let's cover some advanced topics that handle special situations.

Handling Multiple ISO Settings (DSLR Workflows)

Some DSLR astrophotographers shoot at different ISOs for different targets or conditions. For example:

- ISO 1600 for most targets
- ISO 3200 for very faint targets
- ISO 800 for bright targets or moon shots

The challenge: You need separate calibration masters for each ISO.

ISO changes your sensor's gain amplification, which affects:

- Bias offset level
- Dark current level (though not the physical dark current, the amplified readout)
- Noise characteristics

You cannot use an ISO 1600 dark to calibrate ISO 3200 lights. The amplification is different.

Solution: Create a calibration library organized by ISO:

```
/Calibration_Library/
  /ISO800/
    master_bias_ISO800.xisf
    master_dark_180s_ISO800.xisf
    master_flat_L_ISO800.xisf
  /ISO1600/
    master_bias_ISO1600.xisf
    master_dark_180s_ISO1600.xisf
    master_flat_L_ISO1600.xisf
  /ISO3200/
```

WBPP can handle multiple ISOs automatically if you organize files correctly or tag them with metadata.

Temperature-Dependent Darks for Cooled Cameras

Cooled cameras give you precise temperature control, which simplifies dark management. But what temperature should you use?

Common choices:

- **-10°C**: Good balance of dark current reduction and cooler power consumption
- **-15°C**: Better dark current reduction, higher power draw
- **-20°C**: Excellent dark current reduction, maximum power draw
- **-5°C**: Warmer, useful for cameras with limited cooling capacity

My recommendation: Pick a standard temperature (-10°C is popular) and use it consistently. Your dark library then works for all imaging sessions at that temperature.

Temperature matching tolerance:

How close must your lights and darks be in temperature? Generally:

- ±2°C: Excellent (negligible difference)
- ±5°C: Acceptable (slight difference, usually not visible)
- ±10°C: Problematic (visible calibration errors)

Cooled cameras usually maintain ±0.5°C, so temperature matching is automatic.

Defect Maps vs. Master Darks

Some cameras create hot or cold pixels—pixels that are permanently bright or dark. Master darks remove these during calibration. But there's an alternative: defect maps.

Defect Map Concept:

Instead of subtracting dark current, you create a map that identifies bad pixels. During preprocessing, these pixels are ignored and replaced by interpolation from neighbors.

Advantages:

- No dark current subtraction (avoids adding noise)
- Permanent solution (defect map doesn't change)
- Works across all exposure times

Disadvantages:

- Doesn't remove actual dark current (only defects)
- Requires manually creating the map
- Not standard in most workflows

When to use:

Defect maps are most useful for cameras with very low dark current (modern cooled CMOS) where dark current subtraction might add more noise than it removes. For most users, standard dark frames are simpler and work well.

Reusing Calibration Masters

A common question: "How long can I reuse my master calibration frames?"

Bias frames:

- Can be reused indefinitely (months to years)
- Only need new ones if camera settings change (ISO, gain)
- Bias characteristics don't change with time or temperature

Dark frames:

- Can be reused for months if temperature-controlled
- Should remake if:
  - Camera's dark current pattern changes (rare, but happens with sensor aging)
  - Temperature changes (DSLRs)
  - You notice calibration quality degrading
- General recommendation: Remake dark library annually

Flat frames:

- Should be reshot every session or at least when:
    - Filters changed
    - Camera rotated
    - Focus changed significantly
    - New dust appears on sensor (check periodically)
- Cannot be reused across different optical configurations
- Flat characteristics change most frequently

2.6 Validating Your Master Calibration Frames

Before using your masters to calibrate precious light frames, verify they're good. Here's a systematic validation process.

Visual Inspection Checklist

Master Bias:

- [ ] Uniformly gray with subtle, smooth variations
- [ ] No obvious patterns, bands, or gradients
- [ ] Histogram centered at low value with narrow peak
- [ ] Very low noise (smooth appearance)

Master Dark:

- [ ] Similar to bias but with higher signal
- [ ] Hot pixel pattern visible
- [ ] No cosmic ray trails (rejection worked)
- [ ] Smooth except for hot pixels
- [ ] Histogram higher than bias, broader distribution

Master Flat:

- [ ] Obvious vignetting (darkening toward corners)
- [ ] Dust shadows visible (if dust present on sensor)
- [ ] Smooth, free of gradients or patterns
- [ ] High SNR (minimal noise)
- [ ] Histogram peak in mid-tones (40-60%)

Statistical Validation

Use **Process > Image > Statistics** on each master:

Master Bias:

- Mean: Should be low (typically 500-1000 for 16-bit)
- Sigma (std dev): Should be small (<100 for 16-bit)
- Median ≈ Mean (symmetric distribution)

Master Dark:

- Mean: Higher than bias (dark current added)
- Sigma: Higher than bias (more noise from dark current)
- Maximum: Shows brightest hot pixels

Master Flat:

- Mean: Should be in mid-range (20000-40000 for 16-bit)
- Sigma: Reasonable given signal level
- Minimum/Maximum: Should not be clipped (0 or 65535)

Comparative Testing

The best validation: test your calibration on a few light frames.

1. Select 3-5 light frames
2. Calibrate them with your masters
3. Examine the results:
    - Histogram should look reasonable (not clipped)
    - No obvious artifacts or patterns
    - Stars look normal
    - Background appears clean

If calibration makes your lights look worse (too dark, clipped, artifacts), your masters have problems.

Common Problems and Diagnosis

If calibrated lights look too dark:

- Dark frames have too much signal (over-subtraction)
- Check: Are dark and light exposure times correct?
- Check: Are dark and light temperatures matched?
- Solution: Verify metadata, reshoot darks if needed

If calibrated lights have patterns:

- Flats have patterns that got imprinted
- Check: Examine master flat—do you see the same patterns?
- Solution: Reshoot flats with better illumination

If calibrated lights have unusual colors (OSC/DSLR):

- Flats might have color casts
- Check: Is your flat light source color-neutral?
- Solution: Shoot flats with neutral light (dawn/dusk sky)

If calibrated lights still have hot pixels:

- Dark frames didn't capture all hot pixels
- Check: Are some pixels in lights brighter than in darks?
- Solution: Reshoot darks with more frames for better hot pixel coverage

Key Takeaways from Chapter 2:

- Master calibration frames are mathematical tools—they must be accurate
- Bias frames measure electronic baseline (quick to shoot, reusable)
- Dark frames must match exposure time AND temperature of lights
- Flat frames must match optical configuration (focus, filter, rotation) of lights
- Stack 50-100 calibration frames for low-noise masters
- Validate your masters before using them on precious light frames
- Organize masters with clear naming (exposure time, ISO/gain, filter, temperature)
- When in doubt, reshoot—better 30 minutes reshooting than ruined data

CHAPTER 3: IMAGE CALIBRATION AND COSMETIC CORRECTION

Applying Your Calibration Masters

You've invested time creating perfect master calibration frames. Now comes the moment of truth: applying them to your light frames. This is where preprocessing either succeeds or fails. Get the calibration right, and you've removed all the systematic errors from your sensor. Get it wrong, and you've actively damaged your data.

I still remember the sinking feeling when I first calibrated a night's worth of Andromeda Galaxy data and saw the results: clipped stars, mottled backgrounds, and strange dark rings around bright areas. I spent hours troubleshooting before discovering my mistake—a single checkbox in ImageCalibration that inverted how flat fields were applied.

Let me guide you through calibration step-by-step, camera type by camera type, so you avoid the mistakes I made.

3.1 The ImageCalibration Process Explained

ImageCalibration is PixInsight's calibration engine. It takes your raw light frames and your master calibration frames, performs the mathematical corrections, and outputs calibrated light frames ready for further preprocessing.

Understanding the Calibration Equation

Let's revisit the fundamental calibration equation with more precision:

Calibrated = (Light - Dark) / (Flat - Bias)

More explicitly:

Calibrated = [(Light - Dark) / (Flat - Bias)] × MedianFlat

That multiplication by MedianFlat at the end is important—it scales the result back to reasonable values. Without it, dividing by fractions would create huge numbers.

Breaking this down step by step:

1. **Subtract dark from light**: Removes bias offset + dark current + hot pixels
2. **Subtract bias from flat**: Removes bias offset from flat (we don't want to divide by that)
3. **Divide corrected light by corrected flat**: Removes vignetting and dust shadows
4. **Multiply by median flat**: Restores reasonable signal levels

This equation assumes:

- Your dark matches your light in exposure time and temperature
- Your flat matches your light in filter and focus
- All data is linear (not stretched)

Understanding Each Parameter

Let's walk through ImageCalibration's interface systematically.

Process > ImageCalibration > ImageCalibration

When you open the process, you'll see multiple sections. We'll cover each one.

Target Frames Section

This is where you load your light frames—the images you want to calibrate.

- **Add Files**: Click to browse and select your light frames
- **Add Files in Current Folder**: Select entire directory
- **Select All / Invert**: Manage which files are processed
- **Clear**: Remove all files

Important: Your light frames should be RAW, FITS, or XISF files straight from your camera or acquisition software. Don't calibrate already-calibrated or stretched images.

Calibration Section

This is where you tell ImageCalibration which master calibration frames to use.

Master Bias:

- Check "Enable" and select your master bias file
- Use this when you have separate bias and dark frames
- If your darks already include bias (common DSLR workflow), you can disable this

Master Dark:

- Check "Enable" and select your master dark file
- **Critical**: Must match light frame exposure time
- **Optimize checkbox**: Enables dark scaling if exposure times don't match exactly
    - Use this only if necessary—it's better to have matching darks
    - PixInsight scales the dark linearly based on exposure time ratio

Master Flat:

- Check "Enable" and select your master flat file
- **Critical**: Must match light frame filter
- One master flat per filter

Defect List:

- Optional: Upload a text file listing defective pixels
- Most users don't use this—dark frames handle hot pixels

Output Section

Controls what ImageCalibration produces.

Output Directory:

- Where calibrated light frames are saved
- Choose a dedicated folder: /Session/Calibrated/

Output Extension:

- File format: .xisf (recommended) or .fit

Prefix / Postfix:

- Adds text to filename
- Example postfix: _cal produces light_001_cal.xisf
- Helps identify calibration stage

Overwrite Existing Files:

- Uncheck this (safety measure)
- Prevents accidentally overwriting originals

On Error:

- "Continue" is usually best
- If one file fails, process continues with others

Calibration Image:

- For advanced workflows
- Most users leave this disabled

Format Specific Section

If processing DSLR RAW files, this appears.

CFA Pattern:

- Bayer pattern of your sensor
- Usually RGGB, but verify for your camera model
- **Get this wrong and colors will be completely incorrect**

Pedestal:

- Adds a small offset to prevent negative values after calibration
- Default is usually fine (0 or small positive value like 100)
- Used when dark subtraction might create negative pixels

The Pedestal: Why It Exists

When you subtract a dark frame from a light frame, some pixels might go negative. This happens when:

- Dark frame pixel value > light frame pixel value at that position
- Noise variations push subtraction result below zero

Negative pixel values are problematic because:

- Many image formats can't represent them
- Subsequent operations (like division) behave strangely
- You lose information

The pedestal adds a small constant to all pixels after dark subtraction, ensuring everything stays positive. A pedestal of 100 means even pixels that would've been -50 after subtraction become +50 instead.

When to use pedestal:

- DSLR imaging (common to need it)
- When lights and darks are very close in level
- If you see clipping warnings during calibration

When not to use pedestal:

- Cooled camera imaging with good SNR
- When lights are significantly brighter than darks
- Processing narrowband (usually high signal)

Default (0 or small value) works for most scenarios. PixInsight automatically prevents negative values even without explicit pedestal.

3.2 DSLR and OSC Camera Workflows

Color cameras (DSLRs and one-shot color astro cameras) have a Bayer Color Filter Array (CFA) that requires special handling.

Understanding the Bayer CFA

Your color camera sensor doesn't actually capture RGB at each pixel. Instead, it has a mosaic of tiny color filters in a specific pattern:

R G R G R G

G B G B G B

R G R G R G

G B G B G B

This is called a Bayer pattern (specifically RGGB—Red, Green, Green, Blue). Each pixel sees only one color. The camera then "debayers" this pattern, interpolating the missing colors at each pixel to create a full RGB image.

Key insight: Your raw files from color cameras are actually monochrome images with color information encoded in a pattern. They look greenish-tinted when viewed because twice as many pixels are green (human vision is most sensitive to green).

The Debayering Decision: When to Do It

You have two choices:

Option 1: Debayer before calibration

- Calibrate the debayered RGB image
- Simpler workflow
- Slightly less accurate (calibration applied after debayering)

Option 2: Debayer after calibration

- Calibrate the Bayer CFA image
- Debayer the calibrated result
- More accurate (calibration applied to actual sensor data)
- **This is the recommended approach**

Why is debayering after calibration better?

Because your calibration frames (darks and flats) were captured with the Bayer pattern in place. When you calibrate the Bayer CFA image directly, you're subtracting like from like: the red pixels' dark from the red pixels' light, green from green, blue from blue.

If you debayer first, you've interpolated colors, creating synthetic pixel values. Calibrating these interpolated values is mathematically less precise.

DSLR Raw Handling in PixInsight

DSLRs save images in proprietary RAW formats (CR2, NEF, ARW, etc.). PixInsight can read these, but you need correct settings.

Critical Setting: Format Explorer

This controls how PixInsight interprets RAW files.

Edit > Format Explorer > RAW (Digital Camera Raw Formats)

Expand RAW and double-click it. A preferences window opens.

Key settings:

Interpolation:

- Set to: **VNG** or **AHD** (high-quality algorithms)
- This only matters if auto-debayering (which we're disabling)

Import as:

- Set to: **Raw Bayer CFA**
- **Critical**: This prevents automatic debayering
- Your raw files load as greenish monochrome Bayer images

Create Super-pixels:

- Uncheck this
- We want full resolution

White Balance:

- "Camera" is fine (we'll adjust later)

With these settings, when you open a CR2/NEF/ARW file, PixInsight loads it as a Bayer CFA image. You'll see the raw sensor data with a greenish tint.

DSLR Calibration Workflow

Let me walk you through calibrating DSLR light frames step by step.

Step 1: Verify Format Explorer Settings

Confirm RAW is set to import as "Raw Bayer CFA" (see above).

Step 2: Open ImageCalibration

Process > ImageCalibration > ImageCalibration

Step 3: Load Light Frames

Click "Add Files" and select your raw light frames (CR2, NEF, ARW, etc.).

They'll load as Bayer CFA images in the file list.

Step 4: Configure Calibration Masters

Master Bias:

- Enable: Check
- Select your master bias file
- **Important**: Your master bias should also be Bayer CFA (not debayered)

Master Dark:

- Enable: Check
- Select your master dark matching light exposure time and ISO
- **Important**: Master dark should be Bayer CFA

Master Flat:

- Enable: Check
- Select your master flat matching your lights' filter (or no filter for DSLR)
- **Important**: Master flat should be calibrated (bias-subtracted) but not debayered

Wait—should flats be Bayer CFA or RGB?

Answer: Calibrated Bayer CFA. Your flats were shot through the Bayer pattern, so they should remain Bayer CFA through calibration. They'll be debayered along with the lights later.

Step 5: Set Output Options

- **Output Directory**: Choose /Session/Calibrated/
- **Output Extension**: .xisf (recommended for efficiency)
- **Postfix**: _cal
- **Overwrite**: Unchecked

Step 6: Verify CFA Pattern

In the "Format Specific" section, verify:

- **CFA Pattern**: Matches your camera (usually RGGB)

If you're unsure of your camera's pattern:

- Canon: Usually RGGB
- Nikon: Usually RGGB
- Sony: Check your model (often RGGB)
- Fuji X-Trans: Special pattern, requires different handling

Step 7: Execute Calibration

Click the blue checkmark (Execute).

ImageCalibration processes each light frame:

1. Subtracts master dark (removes bias + dark current)
2. Divides by calibrated master flat (removes vignetting + dust)
3. Saves calibrated Bayer CFA images to output directory

Progress: Watch the process console. It'll show each file being processed.

Completion: When done, you'll have calibrated Bayer CFA light frames in your output directory.

Step 8: Verify Calibration Results

Open a calibrated light frame and apply screen stretch (press Ctrl+A to auto-stretch).

What to check:

- Stars should look normal (not clipped or distorted)
- Background should be relatively even (no extreme vignetting)
- No obvious artifacts or patterns
- Histogram should look reasonable (not clipped at either end)

Histogram check:

Process > Image > HistogramTransformation

Look at the histogram. The peak should be in the lower portion (linear data is dark), but there should be signal throughout.

If the histogram is crushed to the left with no data, over-subtraction occurred. If everything is clipped at the right, something went wrong.

Step 9: Save Verification

If results look good, you're ready for debayering. If not, troubleshoot before proceeding (see troubleshooting section later).

Debayering: Converting Bayer CFA to RGB

Now we convert the calibrated Bayer CFA images to full RGB.

Process > Debayer > Debayer

Step 1: Load a Calibrated Light Frame

Open one of your calibrated Bayer CFA images.

Step 2: Configure Debayer Process

Bayer/Mosaic Pattern:

- Should auto-detect from file metadata
- If not, select your pattern (RGGB, etc.)

Debayer Method:

- **VNG**: Fast, good quality (recommended for most)
- **AHD**: Slightly slower, potentially better on fine details
- **SuperPixel**: Fast but reduces resolution (don't use)

Evaluate Noise:

- Check this—gives you noise statistics

Output Mode:

- **Auto**: PixInsight decides (usually fine)
- **Target Image**: Replaces current image with debayered version
- **New Image**: Creates new window with debayered version

Step 3: Apply to One Image

Click the checkmark to debayer the current image.

A new RGB image opens. Apply screen stretch (Ctrl+A) to view it.

What you should see:

- Full color image
- Stars with natural colors (blue-white, orange, red)
- Nebulosity with color (if present)
- Sharp details

Step 4: Batch Debayer All Calibrated Lights

Rather than debayering one at a time, we'll batch process all of them.

Close all open images (to avoid confusion).

Process > Batch Processing > BatchFormatConversion

Wait, that's not Debayer! Correct—BatchFormatConversion can debayer during conversion.

Actually, there's a better approach: use **ImageContainer** with Debayer.

Alternative: Manual Batch Processing

1. Open all calibrated Bayer CFA images: **File > Open**, select all
2. Open Debayer process, configure it
3. Click the triangle at bottom: **New Instance**
4. Drag that instance icon onto each image (tedious for 100+ frames)

Better Alternative: Use WBPP or Manual Script

Honestly, batch debayering is clunky without automation. Most people either:

- Use WBPP (which handles debayering automatically)
- Write a simple script to debayer all files in a folder
- Use ImageContainer (advanced)

For learning purposes: Debayer a few frames manually to verify the process works correctly. For actual processing, consider using WBPP's integrated debayering or processing one filter at a time.

OSC Dedicated Camera Workflow

One-shot color (OSC) dedicated astronomy cameras (ZWO ASI294MC, QHY183C, etc.) work similarly to DSLRs but with key differences:

Differences from DSLR:

- Output FITS files (not proprietary RAW)
- Usually already identified as Bayer CFA in metadata
- No auto-debayering to disable (already disabled)
- More consistent behavior

Workflow:

The process is identical to DSLR, but simpler:

1. Your lights are FITS files with Bayer CFA pattern
2. Load them into ImageCalibration
3. Apply master bias, dark, flat (all Bayer CFA)
4. Calibrate
5. Debayer the calibrated results

The advantage: no Format Explorer settings to worry about. FITS files are explicit about being Bayer CFA.

Common Color Camera Issues

Issue: Colors look wrong after debayering

Cause: Wrong Bayer pattern specified.

Solution:

- Check your camera's actual Bayer pattern (camera specs or documentation)
- Re-debayer with correct pattern
- RGGB is most common, but verify

Issue: Image has color artifacts or mottling

Cause:

- Flats weren't debayered correctly
- Calibration applied to wrong pattern

Solution:

- Verify entire workflow uses consistent CFA pattern
- Remake masters if necessary
- Check that flats were Bayer CFA during calibration

Issue: Stars have colored fringes

Cause:

- This might be chromatic aberration (optical, not calibration)
- Or debayering artifacts on bright stars

Solution:

- If optical: Nothing calibration can fix (handle in post-processing)

- If debayering: Try AHD method instead of VNG

3.3 Monochrome Camera Workflows

Monochrome (mono) cameras are simpler than color cameras because they don't have a Bayer CFA. Every pixel sees the full spectrum of light (or whatever your filter passes).

Advantages of Monochrome Cameras

Why many serious astrophotographers use mono cameras:

1. **No debayering needed**: One less processing step, one less place for errors
2. **Higher resolution**: Every pixel captures data; no interpolation
3. **Better sensitivity**: No Bayer filters blocking light (each pixel sees 3× more photons)
4. **Lower noise**: No color interpolation artifacts
5. **Perfect for narrowband**: Can use extremely narrow filters effectively

Disadvantages:

- Need filters for color imaging (LRGB, narrowband)
- More complex imaging sessions (multiple filter changes)
- Slower to collect full color data (need separate exposures per filter)

Cooled Monochrome Camera Characteristics

Modern cooled mono cameras (ZWO ASI1600MM, QHY268M, etc.) share features:

- **TEC cooling**: Regulated temperature (typically -10°C to -20°C below ambient)
- **No Bayer CFA**: True monochrome sensor
- **Low read noise**: 1-3 electrons typical
- **High QE**: 80%+ quantum efficiency
- **Outputs FITS**: Standard astronomical format

Monochrome Calibration Workflow

The workflow is beautifully simple:

Step 1: Open ImageCalibration

Process > ImageCalibration > ImageCalibration

Step 2: Add Light Frames

Select your mono light frames (FITS or XISF files).

Step 3: Configure Masters

Master Bias:

- Enable and select master bias
- Should match camera gain setting

Master Dark:

- Enable and select master dark
- Must match exposure time and temperature
- For cooled cameras, temperature matching is automatic (set point)

Master Flat:

- Enable and select master flat for the specific filter
- **Critical**: Each filter needs its own flat
- A luminance flat won't work for Ha; an Ha flat won't work for OIII

Step 4: Set Output

- Output directory: /Session/Calibrated/L/ (or /Ha/, /OIII/, etc.)
- Postfix: _cal

Step 5: Execute

Click Execute. Calibration proceeds:

1. Subtract master dark
2. Divide by calibrated master flat
3. Save calibrated mono images

That's it. No debayering, no CFA pattern concerns, no color complications.

Step 6: Verify Results

Open a calibrated frame, apply screen stretch.

Check:

- Stars appear clean and sharp
- Background is even

- No vignetting (flat correction worked)
- No hot pixels (dark subtraction worked)
- Dust donuts removed (flat correction worked)

Filter-Specific Considerations

When imaging with monochrome cameras and filters, you process each filter separately:

LRGB Imaging:

- Calibrate L lights with L flat
- Calibrate R lights with R flat
- Calibrate G lights with G flat
- Calibrate B lights with B flat

Narrowband Imaging:

- Calibrate Ha lights with Ha flat
- Calibrate OIII lights with OIII flat
- Calibrate SII lights with SII flat

Each filter is a completely independent preprocessing path until you combine them for color later.

Why separate flats per filter?

Dust shadows are focus-dependent, and different wavelengths focus at slightly different distances (chromatic focal shift). An Ha flat shows dust at a different apparent size than an OIII flat.

Using the wrong flat creates "dust donuts" in your final image—dark or bright rings that appear/disappear when you switch filters.

Managing Multiple Filter Workflows

With mono cameras and multiple filters, organization is critical.

Directory structure:

```
/Session_2025-01-15_M42/
  /Raw/
    /L/
    /R/
    /G/
    /B/
    /Ha/
    /OIII/
  /Masters/
    master_bias_gain100.xisf
    master_dark_300s_gain100.xisf
    master_flat_L.xisf
    master_flat_R.xisf
    master_flat_G.xisf
    master_flat_B.xisf
    master_flat_Ha.xisf
    master_flat_OIII.xisf
  /Calibrated/
    /L/
    /R/
    /G/
    /B/
    /Ha/
    /OIII/
  /Registered/
    /L/
    /R/
    ...
```

This structure keeps each filter's data separate through preprocessing. They'll come together during color combination later.

3.4 Calibrating Your Light Frames

Let's walk through a complete calibration session with best practices and verification steps.

Pre-Calibration Checklist

Before hitting Execute, verify:

- [ ] All master calibration frames created and validated
- [ ] Master darks match light exposure time
- [ ] Master flats match light filters

- [ ] Master darks match light temperature (cooled cameras) or ambient temp (DSLRs)
- [ ] All masters are in same format (Bayer CFA for color cameras, mono for mono)
- [ ] Output directory exists and has sufficient space
- [ ] You've backed up raw data (calibration is non-destructive, but better safe)

Executing Calibration

Step 1: Configure ImageCalibration

Follow workflows from sections 3.2 or 3.3 depending on camera type.

Step 2: Execute

Click the Execute button (blue checkmark).

What happens:

PixInsight processes each light frame:

1. Opens the raw light
2. Opens master dark
3. Subtracts dark from light
4. Opens master flat
5. Divides corrected light by flat
6. Applies pedestal if configured
7. Saves result to output directory

Progress monitoring:

Watch the Process Console (bottom of screen). You'll see:

Processing image 1 of 150: light_001.cr2

Processing image 2 of 150: light_002.cr2

Time estimate:

- DSLR RAW files: 1-3 seconds per frame
- FITS files: 0.5-2 seconds per frame
- 150 frames: 5-10 minutes typical

If errors occur:

The console shows error messages. Common errors:

- "File not found": Master calibration file path is wrong
- "Incompatible dimensions": Master and light frame sizes don't match
- "Invalid CFA pattern": CFA settings incorrect

Fix the error and retry.

Step 3: Verify Calibration Completed

When finished, the console shows:

150 images processed successfully

0 images failed

Check your output directory—you should have 150 calibrated light frames.

Post-Calibration Verification

This is critical: verify calibration succeeded before proceeding.

Visual Verification:

Open 5-10 random calibrated lights. For each:

1. Apply screen stretch (Ctrl+A)
2. Check overall appearance:
   - Stars look normal (not clipped, not dimmed)
   - Background is relatively even
   - No extreme vignetting
   - No obvious artifacts
3. Zoom to 100% and examine:
   - Star cores sharp
   - No hot pixel trails
   - No dust donuts (or they should be gone if they were in raw)
   - No strange patterns

Statistical Verification:

Open Statistics on several calibrated frames:

Process > Image > Statistics

Check:

- **Mean**: Should be reasonable (not near 0 or saturated)
- **Median**: Close to mean
- **MAD (Median Absolute Deviation)**: Indicates noise level
- **Minimum**: Should be slightly above 0 (pedestal working)
- **Maximum**: Should not be 65535 or 1.0 (not clipped)

Histogram Verification:

Process > Image > HistogramTransformation

Apply to a calibrated frame.

The histogram should:

- Have main peak in lower portion (linear data is dark)
- Show data distributed across range (not all crushed left)
- Have no clipping on either end
- Show smooth distribution (no weird gaps or spikes)

Comparison Test:

Open a raw light and a calibrated light side-by-side.

Apply same screen stretch to both.

Observations:

- Calibrated should have less vignetting
- Calibrated should have no dust donuts
- Calibrated should have no hot pixels (if present in raw)
- Calibrated background should be more even

If calibrated looks worse than raw, calibration failed. Troubleshoot before proceeding.

3.5 Evaluating Calibrated Frames

Let's dive deeper into what to look for in successfully calibrated frames.

What Good Calibration Looks Like

Stars:

- Sharp and round (if raw stars were sharp)
- Natural brightness progression (bright stars bright, faint stars faint)
- No colored halos or fringes (beyond natural chromatic aberration)
- Star cores not clipped (unless they were clipped in raw)

Background:

- Relatively even across frame (flat correction removed vignetting)
- Low-level noise (not amplified by calibration)
- No patterns, bands, or gradients (beyond natural sky gradient)
- No dust shadows or donuts

Histogram:

- Main peak in lower 20-30% (linear data)
- Smooth distribution
- No clipping at 0 or maximum
- Tail extends toward brighter values (star signal)

Overall Impression:

- Image looks similar to raw but cleaner
- No obvious artifacts introduced
- Background more uniform
- Ready for further processing

Common Calibration Problems and Their Appearance

Problem: Stars have dark rings around them

Appearance: Bright stars surrounded by dark halos or rings.

Cause:

- Flats were shot at different focus than lights (dust shadow mismatch)
- Over-subtraction from poorly matched darks

Visual example: Star looks like it has a moat around it.

Problem: Background is too dark or has negative pixels

Appearance: Much darker than raw, histogram crushed to left.

Cause: Over-subtraction from darks

- Dark frames had more signal than lights (temperature mismatch, wrong exposure)
- Pedestal not sufficient to prevent negatives

Problem: Image has vignetting after calibration

Appearance: Corners still dark, not corrected.

Cause:

- Master flat didn't have correct vignetting pattern
- Master flat wasn't applied (disabled in ImageCalibration)
- Flat was inverted (applied backwards)

Problem: Dust donuts visible after calibration

Appearance: Circular or oval dark spots, especially visible in smooth areas.

Cause:

- No flat was applied
- Flat was from wrong filter
- Flat was shot at wrong focus

Problem: Hot pixels still visible

Appearance: Bright pixels scattered across frame.

Cause:

- No dark subtraction
- Dark frames didn't capture all hot pixels (too few darks, cosmic rays)
- Temperature mismatch (darks too cold, didn't have all hot pixels active)

Problem: Image has strange colors (color cameras)

Appearance: Unnatural color casts, magenta or green tints.

Cause:

- Wrong Bayer CFA pattern during debayering
- Flats had color cast that got imprinted
- Calibration masters weren't Bayer CFA

Problem: Horizontal or vertical banding

Appearance: Subtle lines across the frame.

Cause:

- Camera's pattern noise not fully removed by bias/dark
- Need more calibration frames to reduce banding
- Some cameras inherently have banding that's hard to remove

3.6 Cosmetic Correction: Removing Hot and Cold Pixels

Even after dark subtraction, you might have persistent hot or cold pixels. Cosmetic correction is a targeted fix for these.

When to Use Cosmetic Correction

Use cosmetic correction when:

- A few hot pixels remain after calibration
- You have stuck/dead pixels that darks don't fully remove
- You're working with data that lacks dark frames (rare)
- Your camera has specific defects you want to address

Skip cosmetic correction when:

- Your calibration already looks clean
- No obvious hot/cold pixels remain
- You're using modern cooled cameras with excellent darks

- You're in a hurry (it's optional for clean data)

Cosmetic correction is an optional refinement step. Many workflows skip it entirely if calibration was successful.

Understanding CosmeticCorrection Process

Process > CosmeticCorrection > CosmeticCorrection

CosmeticCorrection detects and replaces bad pixels with interpolated values from neighboring good pixels.

Two modes:

Auto-detect mode:

- Process automatically finds hot/cold pixels
- Uses statistical detection
- Easier but less precise

Master dark mode:

- Uses your master dark to identify hot pixels
- More accurate
- Requires a master dark

Using CosmeticCorrection with Auto-Detect

Step 1: Open a Calibrated Light Frame

Load one of your calibrated lights.

Step 2: Open CosmeticCorrection

Process > CosmeticCorrection > CosmeticCorrection

Step 3: Configure Detection

Use auto-detect:

- Check "Auto-detect" checkbox

Detection parameters:

- **Hot pixel sigma**: 5.0 (default)
    - Lower = more aggressive (finds more hot pixels)
    - Higher = more conservative
- **Cold pixel sigma**: 3.0 (default)
    - Similar logic for cold (dark) pixels

These sigma values determine how many standard deviations above/below the local median a pixel must be to qualify as defective.

Step 4: Create Preview to Test

Rather than applying to the whole image, create a preview region:

1. Hold Alt key and press 'N'
2. Draw a box around an area with hot pixels (if you know where they are)
3. Release mouse

A preview rectangle appears.

Step 5: Enable Real-Time Preview

Check "Real-time preview" in CosmeticCorrection.

Adjust the sigma values and watch the preview update. You should see hot pixels being replaced.

Step 6: Verify Settings

Zoom in on the preview. Hot pixels should be gone, replaced by smooth interpolation. If you see:

- Too many pixels changed: Sigma values too low (too aggressive)
- Hot pixels remaining: Sigma values too high (not aggressive enough)

Adjust until just the defective pixels are fixed without affecting good pixels.

Step 7: Apply to Entire Image

Disable preview, click Execute.

CosmeticCorrection processes the image, replacing defective pixels.

Step 8: Verify Results

Zoom to 100% and check hot pixel locations. They should be gone, replaced smoothly.

If results look good, save the image. If not, undo (Edit > Undo) and try different parameters.

Using CosmeticCorrection with Master Dark

This is more accurate than auto-detect because it uses the master dark as a defect map.

Step 1: Open a Calibrated Light

Load calibrated light frame.

Step 2: Open CosmeticCorrection

Step 3: Configure Master Dark Mode

Use auto-detect: Uncheck

Master dark:

- Click "Select File"
- Choose your master dark

Step 4: Set Detection Parameters

Hot dark threshold:

- Pixel value above this in master dark = hot pixel
- Set based on your master dark's statistics
- Typical: 500-1000 above median

Cold dark threshold:

- Pixel value below this in master dark = cold pixel
- Typical: 500-1000 below median

How to find good thresholds:

Open your master dark, check statistics:

- **Median**: Base level
- **MAD (scaled)**: Noise level

Set hot threshold = Median + 5×MAD Set cold threshold = Median - 5×MAD

This identifies pixels that are 5 sigma above/below normal.

Step 5: Apply

Click Execute. CosmeticCorrection uses the master dark to identify and fix defective pixels.

Step 6: Verify

Check that hot pixels are removed without affecting good pixels.

Batch Cosmetic Correction

If you're applying cosmetic correction to all calibrated lights, you need batch processing.

Method 1: Process Instance + Manual Application

1. Configure CosmeticCorrection with optimal settings
2. Create process instance (triangle icon > New Instance)
3. Open all calibrated lights
4. Drag process instance onto each image (tedious)

Method 2: Use ImageContainer (advanced)

Method 3: Let WBPP Handle It (if using WBPP)

WBPP has built-in cosmetic correction that auto-processes all lights.

Honestly, for large datasets, either use WBPP or accept that manual cosmetic correction is tedious. Many people skip it if their darks were good.

When Cosmetic Correction Helps vs. Hurts

Helps when:

- Few isolated hot pixels remain
- Stuck pixels are consistently defective
- Your master dark wasn't perfect

Hurts when:

- Applied too aggressively (removes good signal)
- Threshold set incorrectly (replaces normal pixels)
- Applied to already-clean data (unnecessary processing)

My approach: I check a few calibrated lights. If they look clean, I skip cosmetic correction. If I see obvious hot pixels, I apply it. Don't feel obligated to use every tool—only use what your data needs.

3.7 Troubleshooting Calibration Disasters

Things go wrong. Let's diagnose and fix common calibration failures.

Problem: "My frames are too dark/clipped after calibration"

Symptoms:

- Calibrated lights much darker than raw

- Histogram crushed to left
- Possible negative pixel warnings
- Background nearly black

Diagnosis flowchart:

Step 1: Check your master dark

Open master dark, check statistics:

- If mean/median is extremely high, your dark is overexposed
- If higher than your lights' typical background level, over-subtraction will occur

Step 2: Verify metadata matching

Open a raw light and your master dark. Check FITS headers:

- **EXPTIME**: Must match
- **GAIN** or **ISO**: Must match
- **SET-TEMP** or **CCD-TEMP**: Should be similar (within 5°C)

If any mismatch, that's your problem.

Step 3: Check ImageCalibration settings

- Was "Optimize" enabled for darks? It might be scaling incorrectly
- Try disabling optimization and using exact-match darks

Solutions:

If exposure time mismatch:

- Reshoot darks at correct exposure time
- Or use optimization (less ideal)

If temperature mismatch:

- For DSLRs: Reshoot darks at matching ambient temperature
- For cooled cameras: Verify lights were at expected temperature

If darks are just too hot:

- Darks might have been shot in hot conditions
- Reshoot darks at proper temperature

Workaround:

- Calibrate without darks (use only bias and flat)
- You'll retain hot pixels, but won't over-subtract
- Handle hot pixels with cosmetic correction later

Problem: "Strange patterns appeared after calibration"

Symptoms:

- Gradients, bands, or patterns not present in raw
- Background looks uneven in strange ways
- Artifacts that don't look like astronomical objects

Diagnosis:

Step 1: Examine master flat

Open master flat, apply screen stretch. Do you see:

- Strong gradients? Flat illumination was uneven
- Patterns or rings? Light source created artifacts
- Very noisy appearance? Too few flats or underexposed

If master flat has problems, they'll transfer to calibrated lights.

Step 2: Check flat application

Was the flat actually applied?

- Verify "Enable" was checked for Master Flat in ImageCalibration
- Verify correct flat was selected (right filter)

Step 3: Test without flat

Recalibrate one light frame without the flat (disable flat in ImageCalibration). Does the pattern disappear?

If yes, your flat is the problem.

Solutions:

If flat illumination was bad:

- Reshoot flats with better light source

- Use different flat shooting method (sky, EL panel)

If flat has patterns:

- Change light source
- Add more diffusion
- Shoot more flats to average out patterns

If flat is wrong filter:

- Use correct flat for filter (Ha flat for Ha lights, etc.)

Problem: "Colors look wrong on my calibrated OSC frames"

Symptoms:

- Unnatural color casts (magenta, green)
- Stars wrong colors
- Overall color balance way off

Diagnosis:

Step 1: Check Bayer CFA pattern

Wrong pattern causes completely incorrect colors.

- Verify your camera's actual pattern (camera specs)
- Check ImageCalibration "CFA Pattern" setting
- Try re-debayering with different pattern

Step 2: Check flat frame colors

If your flat light source had a color cast, it gets imprinted.

- Open master flat (before debayering), check if one color channel dominates
- Flats should be relatively neutral

Step 3: Verify calibration masters format

- Were all masters (bias, dark, flat) in Bayer CFA format?
- If any were debayered, you'll get wrong results

Solutions:

If wrong Bayer pattern:

- Re-debayer with correct pattern
- Common patterns: RGGB, BGGR, GRBG, GBRG

If flat color cast:

- Reshoot flats with neutral light source (dawn/dusk sky is best)

If masters were wrong format:

- Remake masters ensuring they're Bayer CFA (not debayered)

Problem: "Calibration made my images worse"

Symptoms:

- Raw lights look better than calibrated
- Artifacts introduced
- Loss of detail or SNR

This should never happen—calibration should always improve images. If it makes things worse, something is fundamentally wrong.

Diagnosis:

Step 1: Verify you have actual problems to fix

Open raw lights. Do they have:

- Hot pixels? (darks needed)
- Vignetting? (flats needed)
- Background gradients from sensor? (calibration might help)

If raw lights are already clean (modern camera, good optics), aggressive calibration might add noise without benefit.

Step 2: Check calibration master quality

Poor quality masters make images worse:

- Noisy masters add noise
- Incorrect masters introduce artifacts

Review master statistics—are they low noise?

Step 3: Test selective calibration

Try calibrating with only:

- Bias + flat (skip darks)
- Dark only (skip flat)

- Flat only (skip bias/dark)

Which combination gives best results?

Solutions:

If masters are low quality:

- Reshoot with more frames
- Ensure proper shooting technique

If calibration isn't needed:

- Some modern cameras with short exposures might not benefit from full calibration
- Consider using just flats (most visually important)

If you're fighting the process:

- Sometimes best approach is no calibration (rare, but possible)
- Focus on good acquisition rather than trying to fix bad data

Key Takeaways from Chapter 3:

- ImageCalibration applies mathematical corrections using your master frames
- Color cameras require careful handling of Bayer CFA pattern—debayer after calibration
- Monochrome cameras have simpler workflows with no debayering
- Verify calibration succeeded before proceeding (histogram, visual check)
- Cosmetic correction is an optional refinement for remaining hot/cold pixels
- When calibration fails, systematically diagnose: check masters, verify metadata, test component by component
- Poor calibration is worse than no calibration—ensure masters are high quality

In Chapter 4, we'll assess the quality of your calibrated frames and select the best ones for integration, using SubframeSelector and manual review techniques.

CHAPTER 4: SUBFRAME SELECTION AND QUALITY ASSESSMENT

Choosing Your Best Data

You've successfully calibrated 150 light frames. They look clean, backgrounds are even, and calibration removed the systematic errors. Now comes a critical question: should you stack all 150 frames, or should you be selective about which frames make it into your final integration?

The answer surprises many beginners: **more frames doesn't always mean better results**. A single out-of-focus frame can soften your entire stack. A frame captured during poor seeing can degrade star shapes. A frame with tracking errors introduces blur. The secret to exceptional astrophotography isn't just collecting lots of data—it's collecting lots of *good* data and being ruthless about rejecting the bad.

This chapter teaches you how to evaluate frame quality systematically and select only your best exposures for integration.

4.1 Why Subframe Selection Matters

The Quality Multiplier Effect

Imagine you have 100 frames with the following quality distribution:

- 70 frames: excellent (sharp focus, good tracking, stable seeing)
- 20 frames: good (slightly soft focus or minor tracking drift)
- 10 frames: poor (out of focus, bad tracking, or terrible seeing)

If you stack all 100 frames, those 10 poor frames drag down your result. They add noise and blur more than they add signal. Your final image quality is limited by your worst frames, not your best ones.

But if you reject the worst 10 and stack only the best 90, you get a sharper, cleaner result despite less total integration time. The math works out because image quality isn't linear—one bad frame damages the stack disproportionately.

Professional astrophotographers routinely reject 20-30% of their frames after quality assessment. Some extremely selective imagers reject 50% or more, keeping only the sharpest frames captured during the steadiest seeing conditions.

What Makes a Frame "Good" or "Bad"?

Several measurable factors determine frame quality:

Focus Quality (FWHM - Full Width Half Maximum)
FWHM measures star sharpness in arc-seconds or pixels. Smaller values = sharper stars. A perfectly focused frame might have FWHM of 2.5 pixels, while a soft frame shows 4.5 pixels. Stacking the soft frame blurs your entire result.

Star Roundness (Eccentricity)
Eccentricity measures how circular stars appear. Values near 0 = round; values near 1 = elongated. Elongated stars indicate tracking errors, wind shake, or other problems. A frame with eccentricity >0.70 probably has issues worth investigating.

Signal-to-Noise Ratio (SNR)
Higher SNR means more signal relative to noise. Frames shot through thin clouds, during bright moonlight, or with poor transparency have lower SNR. These frames contribute less useful data per unit of noise they add.

Star Count
Fewer detected stars might indicate focusing problems, clouds passing through, or other issues reducing the quality of that particular exposure.

Background Level
Unusually bright backgrounds indicate light pollution, moonlight, or twilight encroaching. Dark backgrounds from excellent transparency are preferable.

4.2 Using SubframeSelector

SubframeSelector is PixInsight's automated quality measurement tool. It analyzes each frame and generates quality metrics you can use to approve, weight, or reject frames.

Opening and Configuring SubframeSelector

Process > SubframeSelector > SubframeSelector

The interface has several sections. Let's configure it properly.

Routine Section:

Subframes:
Click "Add Files" and select all your calibrated light frames. SubframeSelector loads them for analysis.

Output Directory:
Choose where measurements will be saved. SubframeSelector creates CSV files with statistics—useful for reviewing later.

Subframe Scale:
This tells SubframeSelector your image scale in arc-seconds per pixel. It's critical for accurate FWHM measurements in physical units.

Calculate your image scale:

Image Scale = (Pixel Size / Focal Length) × 206.265

Example: 3.76μm pixels, 480mm focal length:

(3.76 / 480) × 206.265 = 1.62 arc-seconds per pixel

Enter this value. If you don't know it, SubframeSelector can still work in pixel units, but arc-second measurements are more meaningful.

Camera Gain:
Enter your sensor gain in e-/ADU (electrons per analog-to-digital unit). This helps with noise calculations. Check your camera specifications. Typical values: 0.3-1.0 for modern cameras.

System Geometry Section:

Site:
Your observing location. Used for precise calculations if needed. Optional for basic use.

Focal Length:
Enter your telescope/lens focal length in millimeters.

Pixel Size:
Enter your camera's pixel size in microns.

These inputs ensure accurate measurements.

Expressions Section:

This powerful section defines weighting formulas. For now, use presets:

Weighting Expression:
Select a preset based on your target type:

- **Default**: General purpose
- **Galaxy**: Emphasizes SNR and background
- **Planetary Nebula**: Emphasizes sharpness
- **Star Field**: Balanced approach

These presets create weighting formulas that combine FWHM, eccentricity, and SNR into a single quality score.

Star Detection:

SubframeSelector needs to detect stars to measure them. The defaults usually work, but you might need adjustments for unusual data.

Structure Layers:
Typically 5 layers. This is a multi-scale star detection parameter. Default is fine for most cases.

Noise Layers:
Typically 0-2. Helps reject noise from star detection.

Hot Pixel Filter:
Enabled. Prevents hot pixels from being detected as stars.

Sensitivity:
Default 0.1. Lower = more sensitive (finds fainter stars). Higher = less sensitive (only bright stars).

For normal deep-sky images, defaults work well.

Measuring Your Frames

Once configured, click **Measure**. SubframeSelector analyzes each frame:

1. Detects stars in the frame
2. Measures FWHM, eccentricity, and SNR for each star
3. Calculates frame-wide statistics (median FWHM, median eccentricity, etc.)
4. Computes weight based on your formula
5. Displays results in the table

Progress: Watch the process console. Measuring 100 frames typically takes 2-5 minutes.

Results Table:

When complete, the table shows one row per frame with columns:

- **Index**: Frame number
- **Approved**: Checkbox (initially all checked)
- **Locked**: Checkbox (prevents changes)
- **Weight**: Quality score (higher = better)
- **FWHM**: Median star sharpness
- **Eccentricity**: Median star elongation
- **SNRWeight**: Signal-to-noise based weight
- **Stars**: Number of stars detected
- **Noise**: Estimated noise in frame
- **Other metrics...**

Interpreting SubframeSelector Results

Let's decode what these numbers mean practically.

FWHM (Full Width Half Maximum):

Lower values = sharper stars.

- <2.5 pixels: Excellent (superb seeing/focus)
- 2.5-3.5 pixels: Good (typical conditions)
- 3.5-4.5 pixels: Acceptable (mediocre seeing)
- 4.5 pixels: Poor (bad seeing, poor focus, or problems)

Sort the table by FWHM (click column header). Your best frames are at the top, worst at bottom.

Eccentricity:

Closer to 0 = rounder stars.

- 0-0.30: Excellent (round stars, good tracking)
- 0.30-0.50: Good (slightly elongated)
- 0.50-0.70: Acceptable (noticeable elongation)
- 0.70: Poor (significantly elongated, tracking issues)

High eccentricity might indicate:

- Tracking errors (mount not keeping up)
- Wind shake (mount vibrating)
- Rapid atmosphere changes during exposure
- Dithering artifacts (if dithering aggressively)

SNRWeight:

Higher values = better signal-to-noise.

This combines actual signal level with noise estimation. Frames shot through clouds or during bright conditions score lower.

Weight:

This is the combined quality score using your weighting expression. Higher weight = higher quality frame.

SubframeSelector uses this during integration: high-weight frames contribute more to the final stack than low-weight frames.

4.3 Approving and Rejecting Frames

Now you decide: which frames make the cut?

Setting Approval Thresholds

Click a column header to sort by that metric. For example, sort by FWHM to see sharpest frames first.

Review the distribution:

Look at your FWHM values across all frames:

- Best frame: 2.3 pixels
- Median frame: 3.1 pixels
- Worst frame: 5.2 pixels

Where should you draw the line?

Conservative approach: Reject only the worst outliers (>4.5 pixels FWHM). Keep 90% of frames.

Moderate approach: Reject anything significantly worse than median (>4.0 pixels). Keep 70-80% of frames.

Aggressive approach: Keep only the sharpest 60-70% (reject anything >3.5 pixels). Prioritize quality over quantity.

My recommendation: Start moderate. Reject obvious outliers. As you gain experience, become more selective.

To reject a frame:

Uncheck the "Approved" checkbox for that frame. Rejected frames won't be used in subsequent steps.

Rejection strategy by metric:

FWHM: Reject frames where FWHM is >1.5× your median value.

Eccentricity: Reject frames with eccentricity >0.70 (clearly elongated).

SNRWeight: Look for dramatic outliers (frames 2-3× lower than typical).

Visual inspection: After marking suspects, open and visually inspect them to confirm they deserve rejection.

4.4 Manual Frame Review with Blink

SubframeSelector provides objective metrics, but your eyes catch things algorithms miss: satellites, airplanes, clouds, focusing issues, tracking jumps.

Using Blink for Visual Inspection

Process > Blink > Blink

Blink rapidly displays your frames in sequence, like a flipbook animation. Defects become immediately obvious.

Setup:

1. Click "Add Files" and select your calibrated (and approved) frames
2. Frames load into a list
3. Click "Blink" button

Blink opens a window showing frames rapidly cycling. Watch for:

Satellites/Airplanes:
Bright streaks that appear in one frame and move/disappear in the next. These frames should be rejected.

Clouds:
Sudden dimming or loss of faint stars. The sky looks hazier.

Focus Changes:
Stars suddenly bloat then return to sharpness. Focus drifted during session.

Tracking Jumps:
Star positions suddenly shift significantly. Mount had a tracking glitch.

Wind Shake:
One frame shows elongated stars (wind gust hit during that exposure).

Hot Pixels:
Bright spots that don't move with stars. These should have been removed by darks/cosmetic correction, but if they weren't, note the frames.

To reject a frame: Stop blinking, note the problematic frame number, return to SubframeSelector and uncheck its approval.

Blink is tedious for 150 frames but invaluable for catching issues SubframeSelector misses. Many experienced imagers blink every frame before integration—it takes 10-15 minutes but saves hours of dealing with artifacts later.

4.5 Selecting the Reference Frame

The reference frame is the image that all other frames will be aligned to during registration. Choosing a good reference frame is critical for optimal alignment.

Criteria for the Best Reference

The ideal reference frame has:

1. **Sharpest stars (lowest FWHM):** Makes alignment more precise
2. **Round stars (low eccentricity):** No tracking errors to propagate
3. **Good star distribution:** Stars across the entire frame for alignment
4. **High SNR:** Clear star detection
5. **Typical background:** Not unusually bright or dark

How to select:

Sort SubframeSelector table by Weight (highest first). The top frame is usually your best reference candidate.

Verify the top frame:

Open it, apply screen stretch, examine:

- Stars sharp throughout frame?
- No obvious defects
- Good exposure level?

If yes, this is your reference. If the top frame has issues (perhaps a satellite trail SubframeSelector didn't catch), pick the next highest weight frame.

Mark the reference:

Note the filename. During registration (next chapter), you'll explicitly select this frame as the reference.

SubframeSelector can also auto-select reference for you if you're using WBPP, but knowing how to choose manually gives you control.

Key Takeaways from Chapter 4:

- Frame quality varies—measuring and selecting improves final results
- SubframeSelector provides objective metrics: FWHM (sharpness), eccentricity (roundness), SNR
- Reject frames with FWHM >1.5× median or eccentricity >0.70
- Visual inspection with Blink catches satellites, clouds, and other issues algorithms miss
- Select your sharpest, roundest frame with good star distribution as reference
- Being selective about frame approval improves results more than adding marginal frames
- Don't be afraid to reject 20-30% of frames—quality over quantity

Chapter 5 will cover image registration and alignment, where we'll use your reference frame to perfectly align all approved frames before integration.

CHAPTER 5: IMAGE REGISTRATION AND ALIGNMENT

Getting Your Frames to Match Perfectly

You have 120 approved, high-quality calibrated frames. Each one captures your target at slightly different positions in the frame due to imperfect tracking, atmospheric refraction, and deliberate dithering. Before these frames can be stacked together, they must be precisely aligned so that every star in frame 1 overlaps exactly with the same star in frames 2 through 120.

This alignment process—called registration—is where many preprocessing workflows stumble. Poor registration creates the dreaded "soft stars" that plague amateur astrophotography. Perfect registration produces the razor-sharp results that make images pop. The difference between good and great astrophotography often comes down to registration precision.

5.1 Registration Theory and Importance

Why Perfect Alignment Is Critical

When you stack misaligned frames, each star's light lands on slightly different pixels across the frames. Instead of reinforcing the same pixels (increasing signal), the light spreads across multiple pixels (blurring the star). The mathematical result: your final image has stars that are wider and softer than they should be.

Consider a star that should be 3 pixels wide. If your frames are misaligned by just 0.5 pixels in random directions, that star becomes 4 pixels wide in the stack—a 33% loss in sharpness. If misalignment is 1 pixel, the star bloats to 5 pixels wide. Your expensive optics can't overcome poor registration.

Sub-pixel registration solves this. Instead of aligning to whole pixel boundaries, modern registration algorithms calculate alignment to 0.01 pixel precision or better. A star at position (1024.37, 2048.91) in the reference frame will be aligned to that exact position in every registered frame, not rounded to (1024, 2049).

Understanding Star Matching Algorithms

Registration works by:

1. **Detecting stars** in both the reference frame and the target frame
2. **Creating triangles** connecting groups of three stars
3. **Matching triangles** between frames (unique triangle patterns identify the same stars)
4. **Computing transformation** (translation, rotation, scale) that aligns the frames
5. **Applying transformation** with sub-pixel interpolation

This triangle-matching approach is robust because triangle patterns are scale and rotation invariant. The same three stars form the same triangle shape regardless of how the camera is positioned.

Distortion Correction

Real optical systems introduce distortion—straight lines become slightly curved, especially at frame edges. If your frames show significant distortion (wide-field imaging, fast focal ratios), registration can model and correct this.

Distortion types:

- **Pincushion:** Image compressed toward edges (telephoto lenses)
- **Barrel:** Image expanded toward edges (wide-angle lenses)
- **Irregular:** Complex distortions from optical aberrations

For most astrophotography (moderate focal lengths, quality optics), distortion correction isn't necessary. Enable it only if you notice alignment errors at frame edges despite good central alignment.

5.2 Using StarAlignment Process

StarAlignment is PixInsight's registration workhorse. It handles everything from simple translation-only alignment to complex distortion correction.

Process > StarAlignment > StarAlignment

Reference Image Selection

Critical first step: Load your reference frame.

Click "Reference Image" section, then "From File" button. Browse to your reference frame—the sharpest frame you identified in Chapter 4.

Alternatively, if you have your reference frame already open in PixInsight, select "Active Window" and ensure that window is active.

The reference frame appears as a thumbnail in StarAlignment.

Why reference selection matters:

Every other frame will be aligned to match this reference. If your reference has tracking errors or focus issues, those problems propagate to all registered frames. Always use your best frame as reference.

Target Images

Add Files: Click this and select all your approved calibrated frames (the ones you approved in SubframeSelector).

Important: Don't include the reference frame in this list—it's already specified separately. Including it would register it to itself (unnecessary).

Your target frames appear in a list. These are the frames that will be transformed to match the reference.

Working Mode

Register/Union (default): Aligns frames and crops them all to the largest common area. Recommended for most work.

Register: Aligns frames but doesn't crop. Output frames have different sizes. Useful for mosaics or special workflows.

Use **Register/Union** unless you have a specific reason not to.

Registration Model

This defines what transformations are allowed during alignment:

Translation Only:
Only X/Y position shifts. No rotation, no scale changes, no distortion.

- Use when: Camera wasn't rotated during session, field rotation is negligible
- Fastest processing
- Most rigid (least flexible)

Translation + Rotation:
Allows position shifts and rotation.

- Use when: Field rotation occurred (long sessions, equatorial mounts near poles)
- Most common choice
- Good balance of flexibility and stability

Translation + Rotation + Scale:
Adds scaling capability.

- Use when: Combining data from different instruments with slightly different image scales
- Or atmospheric refraction caused apparent scale changes
- More flexible but can over-fit

Full Projective + Distortion:
Maximum flexibility, models complex optical distortions.

- Use when: Wide-field imaging with significant barrel/pincushion distortion
- Or
- plex
- Can over-fit to noise if not enough stars

My recommendation: Start with **Translation + Rotation**. It handles most real-world sessions (minor field rotation, no instrument changes) without over-fitting. Upgrade to more complex models only if you notice alignment failures.

Star Detection Parameters

StarAlignment must detect stars before it can match them. These parameters control detection sensitivity.

Sensitivity:
Default 0.1. Lower = more sensitive (detects fainter stars). Higher = less sensitive (only bright stars).

For typical deep-sky: 0.1 works well (detects several hundred to thousands of stars).

For sparse star fields: Lower sensitivity (0.05) to detect more stars.

For crowded fields (Milky Way): Higher sensitivity (0.2) to avoid confusion from too many stars.

Peak Response:
Default 0.8. Controls how "star-like" a detection must be to qualify. Higher = stricter (only obvious stars).

Leave at default unless you have issues with false detections.

Maximum Stars:
Limits how many stars are used for matching. Default 2000-5000.

More stars = more accurate matching but slower processing. Default is fine for nearly all cases.

Detection Scales: The Critical Parameter

This is **the most important parameter** in StarAlignment and the one that causes 90% of registration failures.

Detection scales (measured in pixels) defines the size range of stars to detect. Stars outside this range are ignored.

Why it matters:

If your stars are typically 3-5 pixels wide, but detection scales is set to 6-12 pixels, StarAlignment won't detect your stars. Registration fails with "insufficient stars detected" error.

How to set it correctly:

Open your reference frame. Find a typical isolated star (not the brightest, not the faintest—a mid-brightness field star). Zoom to 100%.

Measure the star's diameter in pixels by counting across its profile. It might be 3 pixels, or 5 pixels, or 8 pixels depending on your sampling and seeing.

Set detection scales to bracket this size:

- If stars are ~3 pixels: Use 2-4 or 2-6
- If stars are ~5 pixels: Use 3-7 or 4-8
- If stars are ~8 pixels: Use 5-10 or 6-12

Format: Enter as "minimum-maximum" like 3-7.

Common mistake: Leaving default values (often 6-12 for large stars) when your stars are actually small (3-4 pixels). This guarantees failure.

Verification: After setting detection scales, click the eyeball icon in StarAlignment. This shows detected stars overlaid on your reference. You should see hundreds or thousands of green circles on stars. If you see none or very few, your detection scales are wrong.

5.3 Registration for Different Scenarios

Different imaging situations require different registration approaches.

Single-Session, Single-Instrument (Standard Case)

Scenario: You imaged one target for 3 hours on one night with one telescope and camera. Frames are from a continuous session.

Approach:

- Reference: Your sharpest frame
- Registration model: Translation + Rotation
- Detection scales: Match your star size
- Distortion correction: Disabled

This is the simplest, most common case. Registration should succeed easily with default settings (after adjusting detection scales).

Multi-Session, Same Instrument

Scenario: You imaged across multiple nights (same target, same equipment). Frames might have different focus, field rotation, or atmospheric conditions.

Challenge: Different nights might have different:

- Focus (stars different sizes)
- Field rotation angle
- Seeing quality (star sharpness varies)

Approach:

- Reference: Best frame from best night
- Registration model: Translation + Rotation (handles field rotation differences)
- Detection scales: Wide enough to cover star size variations (e.g., if one night has 3px stars and another has 5px stars, use 2-7)
- Consider registering each night separately, then combining registered masters

Pro tip: If star sizes vary dramatically between sessions (one night superbly focused, another soft), register each session to its own best frame, integrate separately, then integrate the session masters together. This prevents soft frames from one session degrading sharp frames from another.

Dithered Data

Scenario: You dithered between exposures (intentionally shifted telescope position by a few pixels each frame).

Why dither: Dithering moves stars to different pixel positions each frame, allowing stacking to average out hot pixels, bad columns, and other fixed-pattern sensor defects. It also enables drizzle integration for super-resolution.

Challenge: Every frame has different star positions—that's the point!

Approach:

- This is exactly what registration handles—it aligns despite dithering
- Registration model: Translation (+ Rotation if field rotation also present)
- Detection scales: Normal for your star size
- **Enable drizzle data generation** if you plan drizzle integration (see section 5.4)

Registration handles dithered data naturally. The star-matching algorithm doesn't care that stars moved between frames—it finds them and aligns them.

Wide-Field Imaging

Scenario: Short focal length (wide field of view), significant lens distortion.

Challenge: Stars at frame edges don't align well with simple translation/rotation—optical distortion curves the field.

Approach:

- Registration model: Full Projective + Distortion
- Enable distortion correction (see below)
- May need more stars for robust distortion modeling (lower sensitivity)

Distortion correction settings:

In StarAlignment, **Distortion Correction** section:

- **Enable:** Check
- **Order:** Start with 2 (second-order polynomial)
- Higher orders model more complex distortion but risk over-fitting

For most wide-field work, second-order distortion correction handles barrel/pincushion adequately.

5.4 Drizzle Integration Preparation

Drizzle is an advanced stacking algorithm that uses dithered data to achieve resolution beyond your camera's physical pixel sampling. If you dithered during acquisition, drizzle integration can sharpen your final image significantly.

Enabling Drizzle Data Generation

In StarAlignment:

Output Section:

Generate Drizzle Data: Check this box.

Drop Shrink: Default 0.9 (can adjust 0.5-1.0)

- Lower values = more sharpening but more processing time
- 0.9 is a good balance

StarAlignment will generate .xdrz files alongside each registered image. These drizzle data files contain the precise geometric transformation information needed for DrizzleIntegration later.

When to use drizzle:

- You dithered during acquisition (required!)
- Star FWHM is ≥2 pixels (under-sampled optically)
- You want maximum resolution/detail

When to skip drizzle:

- No dithering (drizzle requires dithered data)
- Already well-sampled (FWHM < 2 pixels)
- Speed is more important than ultimate sharpness

Drizzle integration happens after regular integration (Chapter 6). Generate the drizzle data during registration, then decide later whether to use it.

5.5 Executing Registration

Pre-flight check:

- [ ] Reference frame loaded and verified as sharp
- [ ] Target frames loaded (approved frames only)
- [ ] Registration model selected (Translation + Rotation for typical)
- [ ] Detection scales set correctly for your star size
- [ ] Working mode: Register/Union
- [ ] Output directory specified

Click **Execute** (blue checkmark).

What happens:

StarAlignment processes each frame:

1. Detects stars in reference
2. Detects stars in target frame
3. Matches star patterns
4. Computes transformation
5. Applies transformation with sub-pixel interpolation
6. Crops to common area (Union)
7. Saves registered frame

Progress: Watch console. Processing 100 frames typically takes 5-15 minutes depending on frame size and complexity.

Completion: Console reports:

XX images registered successfully

YY images failed

Success rate:

- 100% success: Excellent, all frames aligned
- 95%+ success: Good, a few problematic frames failed
- <90% success: Problem exists, investigate failures

5.6 Troubleshooting Registration Failures

Problem: "No stars detected" or very few stars

Cause: Detection scales wrong.

Solution:

- Measure actual star size in reference frame
- Set detection scales to bracket that size (±2 pixels)
- Use detection preview (eyeball icon) to verify stars are detected

Problem: "Failed to register X images"

Causes:

- Those frames are too different (severe tracking error, clouds, satellite trails)
- Star pattern doesn't match (field rotation too extreme, frame rotated 180°)

Solution:

- Open failed frames individually, inspect them
- If they look bad, reject them (this is what failure detection is for)

- If they look fine, adjust registration model to handle more flexibility

Problem: "Images register but results look wrong"

Cause: Over-fitting—registration found a transformation but it's incorrect.

Solution:

- Reduce model complexity (use Translation instead of Projective)
- Increase detection scales to find more stable star matches
- Verify reference frame is actually sharp and appropriate

Problem: "Registration succeeds but stars are still slightly misaligned at edges"

Cause: Optical distortion not modeled.

Solution:

- Enable distortion correction
- Use higher-order registration model
- Accept that edge stars might never perfectly align (this is normal)

Key Takeaways from Chapter 5:

- Registration aligns frames to sub-pixel precision using star-matching algorithms
- Reference frame should be your sharpest, roundest frame
- Detection scales must match your actual star size (most common failure point)
- Translation + Rotation handles most real-world sessions
- Distortion correction needed only for wide-field or extreme cases
- Generate drizzle data during registration if you dithered (enables super-resolution later)
- Failed registrations often indicate genuinely bad frames worth rejecting
- Perfect registration is critical—soft stars in final image often trace back to registration issues

Chapter 6 will cover image integration (stacking), where your registered frames finally combine into a single master light with dramatically improved signal-to-noise ratio.

CHAPTER 6: IMAGE INTEGRATION (STACKING)

Combining Your Frames into a Master

You've reached the culminating moment of preprocessing. Your frames are calibrated, quality-assessed, and precisely registered. Now they combine into a single master image through integration—the process most people call "stacking."

This is where the magic of astrophotography reveals itself. Individual frames that looked noisy and dim transform into a clean, deep image bursting with detail. Faint nebulosity invisible in single frames emerges from the noise. Star colors become rich and saturated. The transformation feels almost miraculous, but it's pure mathematics: signal adds linearly while noise adds as the square root, giving you genuine improvement in signal-to-noise ratio.

But integration isn't automatic success. The wrong combination method, improper normalization, or poorly chosen rejection algorithm can degrade your results or introduce artifacts. This chapter teaches you how to integrate frames properly, understand rejection algorithms deeply, and troubleshoot integration problems.

6.1 Integration Theory

Why We Stack: Signal-to-Noise Improvement

Every individual exposure contains:

- **Signal**: Photons from your target (the nebula, galaxy, stars)
- **Noise**: Random variations from sensor read noise, dark current, sky background, and photon statistics

The fundamental problem: faint deep-sky objects often have signal barely above noise in individual exposures. A dim nebula might contribute only 50 electrons to a pixel, while noise contributes ±20 electrons. Your signal-to-noise ratio (SNR) is 50/20 = 2.5—marginally detectable.

When you stack N frames, something beautiful happens mathematically:

- Signal adds directly: N frames × 50 electrons = 50N electrons total
- Noise adds as square root: $\sqrt{N}$ × 20 electrons = $20\sqrt{N}$ electrons total

Your stacked SNR becomes: $(50N)/(20\sqrt{N}) = 2.5\sqrt{N}$

Stack 4 frames: SNR improves from 2.5 to 5.0 (2× better) **Stack 16 frames:** SNR improves from 2.5 to 10.0 (4× better) **Stack 100 frames:** SNR improves from 2.5 to 25.0 (10× better)

This is why astrophotographers chase integration time. More frames = more signal above noise = fainter details become visible.

Average vs. Median: The Fundamental Choice

Two basic ways to combine pixel values:

Average (Mean):
Add all values and divide by count: (v1 + v2 + ... + vN)/N

Advantages:

- Statistically optimal for normal distributions
- Best SNR improvement
- Standard approach for most astrophotography

Disadvantages:

- Outliers pull the average (satellites, cosmic rays affect result)
- Requires rejection algorithms to handle outliers

Median:
Sort all values and take the middle one

Advantages:

- Automatically rejects extreme outliers (satellites disappear!)
- No rejection algorithm needed

- Robust to bad data

Disadvantages:

- Slightly worse SNR than average (loses ~20% efficiency)
- Can create "posterization" artifacts with small frame counts
- Discards information from all frames except the middle value

Which to use?

Average with rejection for normal workflows (the standard approach). You get optimal SNR while rejection algorithms handle outliers.

Median only for small frame counts (<10) where rejection algorithms don't work well, or when you have significant outlier contamination that rejection can't handle.

6.2 ImageIntegration Process Deep-Dive

Process > ImageIntegration > ImageIntegration

ImageIntegration is PixInsight's stacking engine. Let's configure it properly.

Image Integration Tab

Add Files: Select all your registered frames (from StarAlignment output directory).

File Format: Your registered frames should all be .xisf or .fit files.

Weights: This critical setting determines how much each frame contributes to the stack.

Weight Options:

Don't Care: All frames weighted equally. Use when you haven't measured frame quality or all frames are equally good.

Exposure Time: Weights by exposure length. Use when combining frames of different durations (180s and 300s mixed).

Noise Evaluation: Weights by measured noise in each frame. PixInsight automatically measures and weights accordingly.

FITS Keyword: Weights based on keywords in FITS headers. Use this when SubframeSelector generated weights—the keyword is typically SSWEIGHT.

Signal: Weights by signal level. Rarely used.

Median: Weights by median pixel value. Also rarely used.

Average SNR: Weights by signal-to-noise ratio calculated from frame statistics.

My recommendation: If you used SubframeSelector, choose **FITS keyword** and specify SSWEIGHT. Otherwise, use **Noise Evaluation** to let PixInsight automatically weight by noise levels.

Why weighting matters: Your sharpest frames with best seeing should contribute more than soft frames with mediocre seeing. Weighting accomplishes this automatically.

Combination Method

Average: Standard choice. Provides optimal SNR when combined with rejection.

Median: Alternative for small datasets or extreme outlier situations.

Minimum/Maximum: Special purposes (creating sky flats, detecting transients). Not for normal integration.

Use Average unless you have specific reasons for median.

Normalization: Matching Background Levels

Your frames might have different background levels due to changing sky conditions, moonlight, or atmospheric transparency variations. Normalization scales frames to match before integration.

Normalization Options:

No Normalization: Assumes all frames have identical background levels. Only use when you're certain this is true (very controlled conditions).

Additive: Adds/subtracts a constant to match backgrounds. Best for most deep-sky work where sky background varies but scene content is identical.

Multiplicative: Scales entire frame by a factor. Use when frames have different exposures or genuine brightness variations across frames.

Adaptive: Advanced normalization handling local variations. Use for complex situations like combining different sessions with different conditions.

My recommendation: Additive for standard workflows. It matches backgrounds without affecting the relative brightness of objects in your frames.

Why normalize? If you don't normalize and frame 1 has background level 1000 while frame 2 has background level 1200, averaging them creates an uneven background in your stack. Normalization fixes this before integration.

Rejection Parameters

This is where integration becomes an art. Rejection algorithms identify and exclude outlier pixels from the combination—removing satellites, cosmic rays, and other transient artifacts.

6.3 Understanding Rejection Algorithms

Rejection algorithms are the key to clean integrations. Let's understand each one deeply.

No Rejection

Simply averages or medians all pixels without checking for outliers.

When to use: Never for astrophotography. Satellites and cosmic rays will contaminate your stack.

Only exception: If you're using median combination (which inherently rejects outliers), you might disable additional rejection.

Percentile Clipping

Rejects a fixed percentage of highest and lowest values at each pixel.

How it works: If you set "percentile low" to 0.2 and "percentile high" to 0.1, the lowest 20% and highest 10% of values are excluded before averaging.

When to use: Small frame counts (3-10 frames) where statistical methods don't work reliably.

Settings:

- Percentile low: 0.1-0.2 (reject bottom 10-20%)
- Percentile high: 0.1-0.2 (reject top 10-20%)

Limitation: Rejects a fixed fraction regardless of whether outliers exist. Can discard good data if outliers aren't present.

Sigma Clipping (Most Common)

Iteratively rejects pixels that deviate too far from the median in standard deviation (sigma) units.

How it works:

1. Calculate median and standard deviation of pixel values
2. Reject values more than N sigmas from median
3. Recalculate median and sigma without outliers
4. Repeat until convergence

Settings:

- **Sigma low:** How many sigmas below median to reject (typically 3-5)
- **Sigma high:** How many sigmas above median to reject (typically 2-3)

Why different values? Astronomical objects are bright outliers above background—you want to reject cosmic rays (high outliers) more aggressively than accidentally rejecting faint object signal (low outliers).

When to use: 10-50 frames. This is the workhorse algorithm for most astrophotography.

Typical settings:

- Sigma low: 4.0 (conservative, protects faint signal)
- Sigma high: 2.5-3.0 (aggressive, removes cosmic rays and satellites)

Iterations: Algorithm runs 2-5 iterations until stable. More iterations = more thorough outlier detection.

Winsorized Sigma Clipping

Improved version of sigma clipping that's more robust to outliers during the sigma calculation itself.

How it works: Instead of rejecting outliers completely, temporarily replaces them with clipped values during sigma calculation. This prevents extreme outliers from inflating the sigma estimate.

When to use: Same as sigma clipping, but more robust. Slightly better performance in most situations.

Settings: Same as regular sigma clipping.

My recommendation: Use Winsorized Sigma Clipping instead of regular sigma clipping—it's strictly better with no downsides.

Linear Fit Clipping

Models each pixel's expected value as a linear function of frame index, then rejects deviations from this model.

How it works: Fits a line through pixel values across frames, rejects points far from the fitted line. This handles situations where sky background changes smoothly across your session (twilight, moonrise, clouds).

When to use:

- Large frame counts (50+ frames)
- Sessions with changing sky conditions (gradual brightness changes)
- Any situation where sigma clipping struggles with gradients

Settings:

- Rejection tolerance: Similar to sigma values, typically 2.5-4.0

Advantages: Excellent at handling sessions where sky background drifted over time. Can model out gradual changes while still rejecting sudden transients (satellites).

My recommendation: For 50+ frames, especially multi-hour sessions, use Linear Fit Clipping. It's more sophisticated and handles real-world conditions better.

Generalized Extreme Studentized Deviate Test (ESD)

Advanced statistical test for detecting multiple outliers in datasets that approximately follow normal distribution.

When to use: Very large frame counts (100+ frames) where you want maximum outlier detection sophistication.

Advantage: Statistically rigorous approach designed specifically for outlier detection in normally-distributed data.

Limitation: Requires substantial frame counts to work properly. Overkill for typical datasets.

Settings: Significance level (typically 0.05) controls how strictly outliers are defined.

6.4 Integration Settings for Different Scenarios

Let me give you specific configurations for common situations.

Standard Session: 30-100 Frames, Normal Conditions

Combination: Average

Normalization: Additive

Weights: FITS keyword (SSWEIGHT) or Noise Evaluation

Rejection: Winsorized Sigma Clipping

- Sigma low: 4.0

- Sigma high: 2.5

- Clip low/high: Both checked

Generate integrated image: Checked

Evaluate noise: Checked

This handles 90% of normal astrophotography sessions.

Small Dataset: 10-20 Frames

Combination: Average

Normalization: Additive

Weights: Noise Evaluation

Rejection: Percentile Clipping

- Percentile low: 0.2

- Percentile high: 0.1

Generate integrated image: Checked

Percentile clipping works better than sigma clipping with small frame counts.

Large Dataset: 100+ Frames, Long Session

Combination: Average

Normalization: Additive

Weights: FITS keyword (SSWEIGHT)

Rejection: Linear Fit Clipping

- Rejection tolerance: 3.0

Generate integrated image: Checked

Evaluate noise: Checked

Generate rejection maps: Checked

Linear fit clipping handles changing conditions across long sessions. Rejection maps let you verify what was rejected.

Satellite-Heavy Session

Combination: Average

Normalization: Additive

Weights: Noise Evaluation

Rejection: Winsorized Sigma Clipping

- Sigma low: 4.0

- Sigma high: 2.0 (more aggressive)

- Clip high pixels: Checked

Generate rejection maps: Checked

Lower sigma high (2.0 instead of 2.5) more aggressively rejects bright outliers like satellites.

6.5 Executing Integration

Pre-integration checklist:

- [ ] All registered frames loaded
- [ ] Weights configured appropriately
- [ ] Combination method: Average
- [ ] Normalization: Additive
- [ ] Rejection algorithm selected and configured
- [ ] Output format selected (.xisf recommended)

Click **Execute.**

What happens: ImageIntegration processes your stack:

1. Opens all frames
2. Normalizes backgrounds if enabled
3. At each pixel location across all frames:
    - Collects N pixel values (N = frame count)
    - Applies rejection algorithm
    - Combines remaining values using average/median

4. Generates final integrated image
5. Optionally generates rejection maps

Processing time: 30-150 frames typically takes 2-10 minutes depending on frame size and rejection complexity.

Completion: A new image window opens—your integrated master light frame!

6.6 Evaluating Your Integration

Your integrated master is still linear (unstretched), so it will look very dark. Apply screen transfer function to view it:

Press Ctrl+A (auto-stretch) to see what you've created.

What to look for:

Stars: Sharp, round, natural colors, no elongation from misalignment.

Background: Even, smooth, low noise compared to individual frames.

Faint details: Nebulosity, galaxies arms, dust lanes now visible that weren't in single frames.

No artifacts: No satellite trails, no cosmic rays, no hot pixels, no walking noise patterns.

Check the rejection maps (if generated):

Rejection maps show which pixels were rejected as outliers. You should see:

- Scattered individual bright pixels (cosmic rays) throughout
- Linear streaks where satellites passed
- Random distribution, not concentrated in specific areas

If rejection maps show large rejected regions or patterns, something went wrong (too-aggressive rejection, misconfigured parameters).

Key Takeaways from Chapter 6:

- Integration combines frames, improving SNR by $\sqrt{N}$
- Average with rejection is standard (optimal SNR + outlier handling)
- Weighting frames by quality improves results
- Additive normalization matches backgrounds before integration
- Winsorized Sigma Clipping handles most situations (10-50 frames)
- Linear Fit Clipping excels with large datasets and changing conditions
- Rejection parameters balance artifact removal vs. preserving real signal
- Generate rejection maps to verify algorithm worked correctly
- Your integrated master is still linear—stretching comes later

Chapter 7 covers WeightedBatchPreprocessing (WBPP), the automated workflow that handles all these steps in a single process—but now you understand what WBPP is doing under the hood.

CHAPTER 7: WEIGHTEDBATCHPREPROCESSING (WBPP)

Mastering PixInsight's Automation Powerhouse

You've learned preprocessing the hard way: manually creating calibration masters, calibrating lights individually, running SubframeSelector, executing StarAlignment, configuring ImageIntegration. It works beautifully—and it takes hours.

Now imagine accomplishing all of that in a single click. Load your raw files, configure a few settings, press Execute, and walk away. Return hours later to find perfectly calibrated, registered, integrated, and weighted master images waiting for you.

This is WeightedBatchPreprocessing (WBPP)—PixInsight's preprocessing automation script. When it works, it feels like magic. When it fails, it feels like a black box you can't debug. This chapter transforms WBPP from mysterious magic into a tool you understand deeply and control completely.

Because you've mastered manual preprocessing in Chapters 2-6, you now know what WBPP does internally. This knowledge makes all the difference between blind hope and confident automation.

7.1 Understanding WBPP's Role

What WBPP Actually Does

WBPP is essentially a wrapper script that executes the processes you learned manually:

Stage 1: Master Calibration Creation

- Stacks bias frames → master bias
- Stacks dark frames → master dark (optionally using master bias)
- Calibrates flat frames (with bias/dark)
- Stacks calibrated flats → master flat

Stage 2: Light Frame Calibration

- Calibrates lights using master bias, dark, and flat
- Applies cosmetic correction (optional)

Stage 3: Quality Assessment

- Runs SubframeSelector on all calibrated lights
- Measures FWHM, eccentricity, SNR
- Calculates weights for each frame
- Optionally approves/rejects frames based on quality

Stage 4: Registration

- Automatically selects best frame as reference (or uses your specified frame)
- Registers all approved frames to reference
- Generates drizzle data (optional)

Stage 5: Integration

- Integrates registered frames using weights
- Applies your chosen combination method and rejection algorithm
- Outputs final integrated master

Stage 6: Post-Integration (Optional)

- Local normalization (advanced)
- Drizzle integration for super-resolution
- Image integration of multiple masters

Each stage uses the same processes you learned manually. WBPP just chains them together automatically with smart defaults.

The Black Box Problem

WBPP's power is also its weakness: when something goes wrong inside this automated chain, diagnosing the problem is difficult. Did calibration fail? Was registration the issue? Did rejection throw away good data?

The solution: Understanding what WBPP does at each stage (which you now do from Chapters 2-6) lets you mentally trace through the workflow. When WBPP fails, you can identify which stage failed and why.

Common failure points:

- **Calibration masters poor quality** → Calibrated lights look wrong
- **Detection scales incorrect** → Registration fails
- **Rejection too aggressive** → All frames rejected during integration
- **Metadata missing** → WBPP can't group frames correctly

7.2 Preparing Your Data for WBPP

WBPP works best when your data is organized and properly tagged with metadata.

File Organization Strategy

WBPP can work two ways:

Method 1: Directory Structure

Organize files by type:

/Session_2025-01-20/

/Bias/

bias_001.fit

bias_002.fit

...

/Darks/

dark_001.fit

dark_002.fit

...

/Flats/

flat_001.fit

flat_002.fit

...

/Lights/

light_001.fit

light_002.fit

...

WBPP can scan directories and automatically classify files by folder name.

Method 2: FITS Metadata

Files can be mixed, but must have correct FITS headers:

- **IMAGETYP** or **FRAME**: 'LIGHT', 'DARK', 'FLAT', or 'BIAS'
- **EXPTIME**: Exposure time in seconds
- **FILTER**: Filter name (for multi-filter workflows)
- **GAIN** or **ISOSPEED**: Camera gain/ISO
- **SET-TEMP** or **CCD-TEMP**: Sensor temperature (for cooled cameras)

Most acquisition software writes these headers automatically. Verify by opening a file and checking FITS headers (**Process > FITS Header**).

Naming Conventions for Smart Recognition

If FITS headers are missing or incomplete, WBPP can parse filenames:

Format: frametype_filter_exposure_other.fit

Examples:

- light_Ha_300s_001.fit → Light frame, Ha filter, 300 second exposure
- dark_300s_001.fit → Dark frame, 300 seconds
- flat_L_001.fit → Flat frame, Luminance filter
- bias_001.fit → Bias frame

WBPP recognizes patterns like:

- Frame type: light, dark, flat, bias (or LIGHT, DARK, etc.)
- Exposure: 300s, 300sec, 300_seconds
- Filter: Ha, OIII, L, R, G, B, etc.

- Binning: BIN1, BIN2, BINNING_1

Pro tip: Even if your acquisition software writes metadata, clear filenames make manual review easier.

7.3 WBPP Interface Tour

Process > Script > Batch Processing > WeightedBatchPreprocessing

The interface has multiple tabs. Let's walk through each systematically.

Bias, Darks, Flats, Lights Tabs

Each tab has identical controls for adding files of that type.

Add Files: Browse and select files manually.

Add Directories: Select entire directories. WBPP scans for recognized files.

Add Custom: Manually specify files with custom parameters (exposure, filter, etc.)—useful when metadata is missing.

Select All / Invert / Remove: Manage file lists.

File Properties: Shows detected metadata (exposure, filter, binning). Verify WBPP recognized your files correctly.

Critical check: After adding files, review the properties. If WBPP shows "Unknown filter" or wrong exposure times, your metadata or filenames need correction.

Calibration Section

Controls how calibration masters are created and applied.

Master Frame Creation:

Generate master frames: Check this—WBPP will create masters from your bias/dark/flat files.

Number of frames: WBPP automatically uses all provided files. You can limit this, but why would you?

Calibration: Check to enable calibration of lights.

Optimize darks: Enables dark scaling for exposure time mismatches. Disable this unless necessary (prefer exact-match darks).

Evaluate noise: Check—provides useful statistics.

Cosmetic Correction: Optionally enables hot/cold pixel removal.

Master Dark Exposure Tolerance: If you have darks with slightly different exposures than lights (299s vs 300s), this tolerance (in seconds) determines if they're considered matching. Default 1-2 seconds works for typical variance.

Calibration Grouping:

WBPP groups frames by:

- Exposure time
- Filter
- Binning
- Gain/ISO (if varying)

Each unique combination gets its own calibration workflow. This is powerful for complex datasets (multiple filters, multiple exposure times) but can be confusing initially.

Verify grouping: WBPP's console output shows which groups it created. If it created 12 groups but you expected 3, something's wrong with metadata.

Registration Section

Controls how frames are aligned.

Reference Frame Selection:

Auto-select: WBPP chooses the best frame automatically (highest weight from SubframeSelector).

Manual: You specify the reference frame explicitly—useful when you've visually identified the perfect frame.

Registration Model: Translation + Rotation (default and recommended for most).

Star Detection:

Detection scales: Defaults to auto-detection, but you can override. If registration fails, this is the first thing to check (see Chapter 5).

Distortion Correction: Disabled by default. Enable only for wide-field imaging with obvious distortion.

Drizzle Integration:

Generate drizzle data: Check if you dithered and want super-resolution capability.

Drop shrink: Default 0.9 is good.

Image Integration Section

Controls how registered frames are stacked.

Subframe Weighting:

Enable weights: Check—this is why it's called "Weighted" BatchPreprocessing!

Weight preset: Choose based on target type:

- **Default**: General purpose
- **Galaxy**: Emphasizes SNR
- **Planetary Nebula**: Emphasizes sharpness
- **Custom**: Define your own formula (advanced)

These presets calculate quality scores combining FWHM, eccentricity, and SNR.

Combination and Normalization:

Combination: Average (standard)

Normalization: Additive (standard for deep-sky)

Rejection Algorithm:

Pixel rejection: Choose algorithm:

- **Winsorized Sigma Clipping**: Most common (10-100 frames)
- **Linear Fit Clipping**: Large datasets (100+ frames)
- **Percentile Clipping**: Small datasets (<10 frames)

Sigma low/high: Default values (4.0/2.5) work well. Adjust if you have issues with over-rejection or remaining artifacts.

Output Section

Output Directory: Where WBPP saves all generated files—masters, calibrated lights, registered lights, integrated masters.

Output Suffix: WBPP automatically organizes output into subdirectories:

- /masters/ - All master calibration frames
- /calibrated/ - Calibrated light frames
- /registered/ - Registered light frames

File Format: .xisf recommended (faster, more efficient than .fit)

7.4 WBPP Workflows for Different Scenarios

Let me give you complete WBPP configurations for common situations.

Single-Night DSLR Session (One Target, No Filter)

Setup:

1. Add 100 bias frames to Bias tab
2. Add 50 dark frames (matching exposure) to Darks tab
3. Add 30 flat frames to Flats tab
4. Add 120 light frames to Lights tab

Configuration:

- Calibration: Generate masters ✓, Calibrate ✓
- Registration: Auto-select reference ✓
- Integration: Enable weights ✓, Default preset
- Rejection: Winsorized Sigma Clipping (4.0/2.5)

Click Run. WBPP processes everything automatically.

Expected output: One integrated master in /output_dir/masters/master_light_BINNING_1.xisf

Multi-Night LRGB Monochrome Session

Setup:

1. Add bias frames (one set, used for all)
2. Add dark frames organized by exposure time (if you shot different exposures per filter)
3. Add flat frames—one set per filter (flat_L, flat_R, flat_G, flat_B)
4. Add light frames—WBPP will group by filter automatically

Configuration:

- Calibration: Generate masters ✓, Calibrate ✓
- Calibration grouping: By filter ✓
- Registration: Auto-select per group ✓
- Integration: Enable weights ✓

Expected output: Four integrated masters:

- master_light_L_BINNING_1.xisf
- master_light_R_BINNING_1.xisf
- master_light_G_BINNING_1.xisf
- master_light_B_BINNING_1.xisf

WBPP automatically processes each filter independently with correct calibration masters.

Narrowband Ha/OIII/SII Session

Setup: Similar to LRGB—separate flats per filter, lights grouped by filter.

Configuration: Identical to LRGB workflow.

Expected output: Three masters (Ha, OIII, SII), each properly calibrated with its matching flat.

Mixed Exposure Times (HDR Workflow)

Scenario: You shot 60×180s exposures and 40×300s exposures of the same target.

Setup: Add all lights together—WBPP detects different exposures.

Configuration:

- Calibration: Need matching darks for each exposure (dark_180s and dark_300s)
- Integration: WBPP creates separate masters for each exposure duration

Expected output: Two masters:

- master_light_180s_BINNING_1.xisf
- master_light_300s_BINNING_1.xisf

You'll manually integrate these together later for HDR.

7.5 Reading WBPP Output and Reports

WBPP generates extensive console output and reports.

Understanding the Console Log

WBPP's console shows each stage:

*** Creating master bias...

Integration: 100 frames, Average combination

Rejection: Percentile clipping

Output: master_bias.xisf

*** Creating master dark...

Integration: 50 frames, Average combination

Output: master_dark_300s.xisf

*** Calibrating light frames...

Processing: light_001.fit

Processing: light_002.fit

120 frames calibrated successfully

*** Measuring subframe quality...

[SubframeSelector output]

*** Registering frames...

Reference: light_037_cal.xisf (best weight)

Processing: light_001_cal.xisf [OK]

Processing: light_002_cal.xisf [OK]

118 frames registered successfully

2 frames failed registration

*** Integrating registered frames...

Integration: 118 frames, Average combination

Weights: SSWEIGHT keyword

Rejection: Winsorized Sigma Clipping

Output: master_light_BINNING_1.xisf

What to check:

- Master frame counts (verify expected numbers)
- Calibration success rate (should be 100%)
- Registration success rate (95%+ is good, 100% is excellent)
- Integration frame count (approved frames only)

Smart Report

After completion, WBPP displays a Smart Report summarizing:

- Files processed by type
- Master frames created
- Registration statistics (success rate, reference frame)
- Integration statistics (final frame count, rejection counts)
- Output file locations

Save this report if you had issues—it's essential for troubleshooting.

7.6 When WBPP Fails: Diagnosis and Solutions

"All frames rejected during integration"

Symptom: Console shows "0 frames approved" or "All frames rejected."

Diagnosis: Either SubframeSelector rejected everything, or calibration produced unusable frames.

Solution:

1. Check calibrated frames manually—open a few, apply stretch. Do they look correct?
2. If calibrated frames look wrong → Master calibration issue (wrong temperature, wrong exposure)
3. If calibrated frames look fine → SubframeSelector was too strict or broke. Disable subframe weighting.

"Registration failed for most images"

Symptom: 50% or more frames fail registration.

Cause: Detection scales wrong for your star size.

Solution:

1. Open a calibrated frame, measure star FWHM
2. Override WBPP's auto-detection: Set detection scales manually to bracket your star size
3. Re-run WBPP (you can start from registration stage if calibration succeeded)

"Calibrated frames look terrible"

Symptom: After WBPP completes, calibrated frames are clipped, too dark, or have artifacts.

Cause: Master calibration frames are wrong (temperature mismatch, exposure mismatch).

Solution:

1. Check WBPP's master frames in /output_dir/masters/
2. Verify master dark exposure matches lights
3. Check FITS headers on masters and lights—do exposure times match?
4. If mismatch found, fix and re-run WBPP

"Process won't complete / hangs"

Symptom: WBPP runs for hours with no progress or gets stuck.

Cause: Usually memory exhaustion or processing very large files.

Solution:

1. Check memory usage—are you near your RAM limit?
2. Convert RAW to XISF first (more memory-efficient)
3. Process in smaller batches (split lights into groups)
4. Close other applications
5. Consider processing filters separately rather than all at once

Key Takeaways from Chapter 7:

- WBPP automates the entire preprocessing pipeline you learned in Chapters 2-6
- Proper file organization and metadata make WBPP reliable
- WBPP automatically groups frames by filter, exposure, binning, and gain
- Understanding manual workflow lets you diagnose WBPP failures
- Detection scales are the #1 cause of registration failures—override auto-detection if needed
- WBPP console log reveals exactly what's happening at each stage
- Start with simple single-filter sessions, then graduate to complex multi-filter workflows
- When WBPP fails, return to manual processing for that problematic stage

Chapter 8 covers multi-session and complex data integration—handling those situations where simple single-session workflows aren't enough.

CHAPTER 8: MULTI-SESSION AND COMPLEX DATA INTEGRATION

When One Night Isn't Enough

Clear nights are precious. You captured three hours on M31 last month under excellent conditions. This month, you managed two more hours before clouds rolled in. Next month, perhaps another hour or two. Separately, these sessions are good. Combined, they could be spectacular—if you can integrate data from different nights successfully.

Or perhaps you're combining data from different telescopes: your wide-field refractor and your friend's longer focal length SCT. Or mixing 180-second exposures from earlier in the session with 300-second exposures from later. Or dealing with a meridian flip that rotated your field mid-session.

These complex integration scenarios challenge even experienced astrophotographers. The frames don't align perfectly, backgrounds differ between sessions, seeing conditions varied, and equipment differences introduce subtle incompatibilities. Yet mastering multi-session integration unlocks your ability to build deep images over weeks or months, combining every photon you've collected regardless of when or how you captured it.

This chapter teaches you how to handle these challenging integration scenarios successfully.

8.1 The Multi-Session Challenge

What Makes Multi-Session Integration Difficult

When combining data from different nights or equipment, you face challenges that don't exist in single-session workflows:

Different Seeing Conditions:
Night 1 might have 2.5-pixel FWHM (excellent seeing). Night 2 shows 4.5-pixel FWHM (mediocre seeing). Stacking them together means the sharper frames get blurred by the softer ones—unless you weight appropriately.

Focus Variations:
Even with careful focusing each night, slight variations occur. Perfect focus one night, slightly off another. Star profiles don't quite match, making integration challenging.

Field Rotation Differences:
Unless you image at the same hour angle each session, field rotation differs. One session at 2 hours past meridian flip shows different rotation than a session at 4 hours before flip. Registration handles this, but it adds complexity.

Background Level Variations:
Moon phase, light pollution, atmospheric transparency—all affect background levels. Night 1 might have dark backgrounds (new moon, excellent transparency). Night 2 shows brighter backgrounds (thin clouds, worse seeing). Normalization must handle these differences.

Temperature Changes (Seasonal Imaging):
Imaging the same target across months means different sensor temperatures (for cooled cameras) or ambient temperatures (for DSLRs). While calibration handles this, ensuring darks match becomes more complex.

Equipment Changes:
Accidentally bumped collimation between sessions? Camera rotated slightly? Filter wheel position changed? These subtle differences compound integration challenges.

8.2 Combining Different Exposure Times

The HDR Challenge

You have 60 frames at 180 seconds (3 hours total) and 40 frames at 300 seconds (3.3 hours total). Why different exposures? Perhaps you started conservatively with shorter exposures, gained confidence, and switched to longer ones. Or you targeted bright core regions with shorter exposures and faint outer regions with longer ones.

Can you integrate these together? Yes, but carefully.

Approach 1: Weight by Exposure Time

Process:

1. Register all frames (both 180s and 300s) to the same reference
2. Use ImageIntegration with **Weights: Exposure Time**
3. PixInsight automatically weights 300s frames 1.67× higher than 180s frames

This approach treats different exposures as having different "quality"—longer exposures collected more signal, so they contribute more to the stack.

When this works well:

- All frames have similar quality besides exposure length
- No significant differences in seeing, focus, or conditions between exposure groups
- You want a single integrated master combining all data

Configuration:

ImageIntegration:

Add files: All registered frames (180s + 300s)

Combination: Average

Normalization: Additive

Weights: Exposure Time

Rejection: Winsorized Sigma Clipping

PixInsight reads exposure time from FITS headers and weights automatically.

Approach 2: Integrate Separately, Then Combine

Process:

1. Register all frames to common reference
2. Integrate 180s frames separately → master_light_180s
3. Integrate 300s frames separately → master_light_300s
4. Integrate the two masters together with exposure-time weighting

This approach gives you more control and allows different rejection parameters for each group.

When this works better:

- Exposure groups have different characteristics (one group significantly softer)
- You shot them on different nights with different conditions
- You want to verify each group individually before combining

Configuration for final integration of masters:

ImageIntegration:

Add files: master_light_180s.xisf, master_light_300s.xisf

Combination: Average

Normalization: Additive (or Adaptive if backgrounds differ significantly)

Weights: Exposure Time

Rejection: None (only 2 images, no outliers to reject)

HDR for Bright Core + Faint Outer Regions

Some targets (bright galaxies, nebulae with bright cores) benefit from HDR: short exposures capture core detail without saturation, long exposures reveal faint outer structure.

Process:

1. Integrate short exposures → master_short (captures core)
2. Integrate long exposures → master_long (captures outer regions)
3. In post-processing, blend them using masks: long exposure for most of image, short exposure only for saturated core regions

This is technically post-processing (beyond preprocessing scope), but the preprocessing setup requires separate integration of each exposure group.

8.3 Handling Meridian Flips

What Happens During a Meridian Flip

German Equatorial Mounts (GEMs) flip when crossing the meridian—the mount physically rotates 180° to continue tracking. This causes:

Field Rotation:
The entire field rotates 180° instantly. Registration must handle this large rotation.

Potential Focus Shift:
Mechanical flexure during the flip might slightly change focus.

Star Profile Changes:
Different mount/telescope position might show different star shapes (aberrations that vary by orientation).

Guiding Changes:
Guiding behavior might differ on opposite sides of the mount.

When to Process Together vs. Separately

Process together when:

- Registration successfully handles the 180° rotation (it usually does)
- No obvious quality differences between pre-flip and post-flip frames
- Focus stayed good throughout
- Total frame count is reasonable (combined dataset isn't overwhelming)

Process separately when:

- Registration fails (can't handle the rotation—rare but possible)
- Focus shifted during flip (one set is noticeably softer)
- Star profiles look significantly different (aberrations changed)
- You want to compare quality between sides of meridian

Processing Together Approach

Process:

1. Run SubframeSelector on all frames (pre-flip and post-flip together)
2. Select best frame as reference (might be pre-flip or post-flip)
3. Use StarAlignment with **Registration Model: Translation + Rotation**
   - This handles the 180° flip naturally
4. Integrate all registered frames together

Critical setting: Registration model MUST include rotation. Translation-only won't work—the flip is a 180° rotation that must be corrected.

Configuration:

StarAlignment:

Reference: Best frame (either pre-flip or post-flip)

Target images: All frames

Registration model: Translation + Rotation

Working mode: Register/Union

StarAlignment detects the 180° rotation and aligns correctly.

Processing Separately Approach

Process:

1. Separate frames into pre-flip and post-flip groups
2. Process each group independently through registration and integration
3. Result: master_light_preflip.xisf and master_light_postflip.xisf
4. Register these two masters to each other
5. Integrate them together

This gives you more control and allows you to see if one side of the meridian produced better data.

When to use: If you suspect quality differences or registration struggles with the full dataset.

8.4 Multi-Night Integration

Challenge: Accumulating Signal Over Multiple Nights

You're imaging a faint galaxy that requires 20+ hours of exposure. No single night provides that much darkness, so you image across multiple nights—perhaps even multiple months.

Issues to solve:

- Different seeing each night (some nights excellent, others mediocre)
- Different focus accuracy
- Different atmospheric conditions
- Possibly different moon phases affecting background
- Long-term equipment drift (focus, collimation)

Strategy 1: Integrate Each Night Separately First

Recommended approach for multi-night campaigns:

Process:

1. **Night 1:** Process completely → master_light_night1.xisf
2. **Night 2:** Process completely → master_light_night2.xisf
3. **Night 3:** Process completely → master_light_night3.xisf
4. **Final step:** Register the three masters to each other → Integrate together

Why this works better:

- Each night's data is optimized independently (best rejection parameters for that night)
- You can evaluate each night's quality before combining
- Poor nights can be downweighted or excluded entirely
- Different seeing conditions don't cross-contaminate frames

Process for combining masters:

Open StarAlignment:

Reference: Best night's master (choose the sharpest)

Target images: Other nights' masters

Registration model: Translation + Rotation (handles any rotation differences)

Execute registration

Then ImageIntegration:

Add files: All registered night masters

Combination: Average

Normalization: Adaptive (handles background differences between nights)

Weights: Noise Evaluation (gives better nights more influence)

Rejection: None (only a few images, no outliers)

Strategy 2: Integrate All Frames Together

Alternative approach: Throw all frames from all nights into one giant integration.

Process:

1. Calibrate all frames (each night with its appropriate calibration masters)
2. Run SubframeSelector on the entire dataset
3. Register everything to a single reference (best frame across all nights)
4. Integrate with quality weighting

When this works:

- All nights had similar quality
- No significant equipment changes between nights
- You have a powerful computer (processing 300+ frames at once)

Challenge: Poor nights drag down good nights unless weighting is strong enough. A 100-frame session with 2.5px FWHM combined with a 50-frame session with 4.5px FWHM will be softer than the good session alone—unless you aggressively weight by quality.

Configuration:

SubframeSelector: Measure all frames, calculate weights

StarAlignment: Reference = absolute best frame

ImageIntegration:

Weights: FITS keyword (SSWEIGHT from SubframeSelector)

This ensures sharp frames dominate

8.5 Combining Data from Different Instruments

The Scale Matching Challenge

You have:

- 50 frames from your 480mm focal length refractor (1.62 arc-seconds/pixel)
- 30 frames from your friend's 1200mm focal length SCT (0.65 arc-seconds/pixel)

Both targeted the same object. Can you combine them?

Technical challenge: Different image scales mean stars are different sizes in pixels. A 3-pixel star in one dataset is a 7.5-pixel star in the other (480/1200 = 0.4×, so 3/0.4 = 7.5).

When Combination Makes Sense

Combine when:

- Image scales are within 2× of each other (one is 1.5"/pixel, other is 3.0"/pixel)
- Field of view overlaps significantly
- Quality is comparable (both nights had similar seeing in arc-seconds)

Don't combine when:

- Scales differ by 3× or more (too different to reconcile)
- One dataset is dramatically better quality (just use the better one)
- Fields don't overlap well

Process for Different Scales

Approach:

1. Process each dataset completely and independently → two separate masters
2. **Resample** one master to match the other's scale (using ImageResample or Registration with scale adjustment)
3. Register the resampled master to the other
4. Integrate the two masters

Which one to resample? Generally, resample the higher resolution (longer focal length) down to match the lower resolution. This avoids creating artificial resolution where none existed.

Configuration:

After creating both masters:

1. Resample the 1200mm master:

Process > Geometry > Resample

Scale factor: 0.4 (matches 480mm scale)

Interpolation: Lanczos-4 (high quality)

2. Register resampled master to 480mm master:

StarAlignment with 480mm master as reference

3. Integrate:

ImageIntegration of both registered masters

Weight appropriately (consider exposure time, quality)

8.6 Using LocalNormalization for Challenging Data

When Standard Normalization Fails

Sometimes frames have background variations that additive or multiplicative normalization can't fix:

- One corner was affected by thin clouds
- Gradient changed orientation between sessions
- Moon position created different light pollution gradients

LocalNormalization handles spatially-varying background differences.

How LocalNormalization Works

Instead of globally adjusting the entire frame, LocalNormalization models background variations locally across the frame. It identifies low-frequency background components and normalizes them region-by-region.

Process > LocalNormalization > LocalNormalization

When to use:

- Multi-night data with different gradients
- Sessions with changing moon position
- Combining data shot from different locations (different horizon light pollution)

Configuration:

Reference: Best frame (or average of all frames)

Target images: All other frames

Scale: 256 (default, covers typical gradient scales)

Execute

LocalNormalization creates normalized versions of all frames that should integrate more cleanly.

Then integrate the locally normalized frames with standard ImageIntegration (no additional normalization needed—already done).

8.7 Multi-Filter Integration Strategies

LRGB Workflow

You've processed each filter independently:

- master_light_L.xisf (luminance)
- master_light_R.xisf (red)
- master_light_G.xisf (green)
- master_light_B.xisf (blue)

These don't get integrated together (they're different spectral data). Instead, they're combined in post-processing using LRGB combination techniques.

Preprocessing's job: Ensure each master is the best possible quality for its filter. Integration happens per-filter, not across filters.

Narrowband Workflow

Similarly, Ha/OIII/SII are processed independently:

- master_light_Ha.xisf
- master_light_OIII.xisf
- master_light_SII.xisf

Later combined into false-color images (SHO, HOO palettes) during post-processing.

Preprocessing's job: Perfect integration of each filter's data independently.

Synthetic Luminance

Advanced technique: combine Ha+OIII (or all three narrowband channels) into a synthetic luminance for use in LRGB combination.

Process:

1. Integrate Ha, OIII, SII independently → three masters
2. **After stretching** (post-processing), combine them:
3. PixelMath: Synthetic_L = 0.5*Ha + 0.35*OIII + 0.15*SII
4. Use this synthetic luminance with RGB color data

This is post-processing territory, but worth mentioning as a goal of narrowband preprocessing.

Key Takeaways from Chapter 8:

- Multi-session integration requires careful quality matching—weight by session quality
- Different exposure times: integrate separately, then combine with exposure-time weighting
- Meridian flips: use Translation + Rotation registration model to handle 180° rotation
- Multi-night campaigns: integrate each night separately, then combine night masters
- Different instruments: possible if scales are similar, resample to match
- LocalNormalization handles spatially-varying background differences between sessions
- Per-filter integration for LRGB and narrowband—combine filters in post-processing
- Always prioritize quality over quantity—don't let poor sessions drag down excellent ones

Chapter 9 covers narrowband preprocessing workflows specifically—the special considerations for Ha, OIII, and SII imaging that differ from broadband processing.

CHAPTER 9: NARROWBAND PREPROCESSING WORKFLOWS

Special Considerations for Ha, OIII, and SII

Narrowband imaging feels like unlocking astrophotography's secret level. Suddenly, light pollution doesn't matter. You're capturing photons from specific emission lines—hydrogen-alpha at 656nm, doubly-ionized oxygen at 496/501nm, ionized sulfur at 672nm—while rejecting nearly everything else. The city lights that plagued your broadband images? Blocked. The moon? Barely an issue. You're imaging the universe in a language that cuts through the noise.

But narrowband comes with its own challenges. Signals are weak—often dramatically weaker than broadband. A 5-minute broadband exposure might require 20 minutes in narrowband to achieve similar SNR. Integration times stretch into dozens of hours for faint targets. And each filter requires completely independent preprocessing before you can combine them into those stunning false-color images you've admired online.

This chapter teaches you how to preprocess narrowband data effectively, handle the unique challenges of weak signals and long exposures, and prepare your Ha, OIII, and SII masters for color combination.

9.1 Narrowband Imaging Overview

Why Narrowband Is Different

Broadband filters (RGB, Luminance, or no filter) pass hundreds of nanometers of spectrum. Narrowband filters pass only 3-7nm bands centered on specific emission lines. This extreme selectivity creates fundamental differences:

Signal Characteristics:

- **Much weaker:** Blocking 99% of the spectrum means far fewer photons
- **Emission-line dependent:** Bright in emission nebulae, absent in reflection nebulae or galaxies
- **Filter-specific intensity:** Ha is typically strongest, SII weakest, OIII variable

Exposure Requirements:

- **Longer individual exposures:** 10-20 minutes common (vs. 3-5 minutes broadband)
- **More total integration:** 5-10+ hours per filter not unusual
- **Better tracking required:** Long exposures demand excellent guiding

Calibration Considerations:

- **Dark current matters more:** Long exposures accumulate substantial dark current
- **Temperature control critical:** With cooled cameras, maintaining stable temperature is essential
- **Flats absolutely necessary:** Dust donuts are highly visible in narrowband's cleaner backgrounds

Filter Characteristics and Bandpasses

Hydrogen-Alpha (Ha - 656nm):

- **Bandpass:** 3nm, 6nm, or 7nm typical
- **Signal:** Usually strongest of the three
- **Emission source:** Ionized hydrogen (HII regions, planetary nebulae, supernova remnants)
- **Visual appearance:** Red/magenta in natural color

Oxygen-III (OIII - 496nm and 501nm):

- **Bandpass:** 3nm or 6nm, centered on doublet
- **Signal:** Moderate to strong, highly target-dependent
- **Emission source:** Doubly-ionized oxygen (high-excitation regions)
- **Visual appearance:** Blue-green/cyan in natural color

Sulfur-II (SII - 672nm):

- **Bandpass:** 3nm or 6nm
- **Signal:** Typically weakest of the three

- **Emission source:** Ionized sulfur (lower-excitation regions than OIII)
- **Visual appearance:** Deep red, very close to Ha wavelength

Bandpass choice (3nm vs. 6nm):

- **3nm:** Extreme light pollution rejection, but dimmer signal (longer exposures needed)
- **6nm:** More signal (faster), but less light pollution rejection
- **7nm:** Even faster, often used for Ha where signal is abundant

Most astrophotographers use 6nm as a good compromise.

9.2 Camera Considerations for Narrowband

Monochrome Camera Advantages

Narrowband is where monochrome cameras truly shine:

Full Resolution:
Every pixel captures data. No Bayer matrix wastes 2/3 of the photons.

Better Sensitivity:
Without Bayer filters, each pixel sees roughly 3× more light. For already-weak narrowband signals, this is crucial.

Per-Filter Optimization:
Each filter gets independent exposure times, gain settings, and processing. You can shoot Ha for 10 minutes, OIII for 15 minutes, and SII for 20 minutes—whatever each filter needs.

Cleaner Color Separation:
No color contamination between channels. Ha is pure Ha, not Ha-plus-red-Bayer-filtered-light.

OSC Cameras with Narrowband: The Compromise

One-shot color (OSC) cameras can use narrowband filters, but with limitations:

Dual-Band Filters (Ha+OIII):
Pass two emission lines simultaneously. The Bayer matrix separates them: Ha lands mostly on red pixels, OIII mostly on blue pixels. Green pixels get a mix.

Advantages:

- Faster (capturing two channels simultaneously)
- Simpler workflow (one exposure set instead of two)
- Color cameras become viable for narrowband

Disadvantages:

- Channel separation isn't perfect (crosstalk between Ha and OIII)
- No SII capability (can't do full SHO palette)
- Bayer matrix still wastes ~2/3 of photons per channel
- More complex post-processing to separate channels

Tri-Band Filters (Ha+OIII+Hbeta):
Add hydrogen-beta (486nm) for additional detail, but Hbeta is very weak in most targets.

Sensor Sensitivity to Narrowband Wavelengths

Not all sensors respond equally to narrowband wavelengths:

Front-Illuminated Sensors:
Generally good Ha sensitivity, excellent OIII sensitivity (blue-sensitive), good SII sensitivity.

Back-Illuminated Sensors:
Excellent sensitivity across all wavelengths, but can have challenges with OIII (particularly older designs with UV/IR filters).

Modified DSLRs:
Removing the IR-cut filter improves Ha sensitivity dramatically (stock DSLRs block ~75% of Ha). Less important for OIII and SII.

Check your camera's quantum efficiency curve at your filter wavelengths before committing to narrowband. Some cameras have >80% QE at Ha but only 40% at OIII—you'll need 2× longer exposures for OIII to match SNR.

9.3 Calibration for Narrowband

Dark Frames for Long Exposures

Narrowband's long exposures (10-20+ minutes) make dark current significant even for cooled cameras.

Critical factors:

Temperature Matching:
With exposures this long, temperature must match precisely. Even 2-3°C difference creates visible calibration errors. For cooled cameras, use the exact set temperature. For DSLRs, you're at the mercy of ambient temperature—shoot darks immediately before/after lights.

Amp Glow Management:
Long exposures reveal amp glow (electronic glow in sensor corners) more prominently. Your darks must capture this accurately to subtract it.

Cosmic Ray Frequency:
A 15-minute exposure likely has 5-10 cosmic ray hits. Your dark frames will too. Stack 50-100 darks with aggressive rejection (Sigma Clipping at 3.0/2.0) to remove these outliers while preserving the dark current pattern.

Hot Pixel Accumulation:
Long exposures make hot pixels very bright. Ensure your master dark captures all hot pixels by stacking many frames with rejection.

Flat Field Critical Importance

Narrowband's clean, dark backgrounds make dust shadows glaringly obvious. What was barely visible in broadband becomes a constellation of donuts in narrowband.

Dust donut appearance in narrowband:

- **Ha:** Slightly soft-edged donuts (red light, larger diffraction)
- **OIII:** Sharper donuts (blue light, smaller diffraction)
- **SII:** Very similar to Ha (wavelengths nearly identical)

Flat shooting for narrowband:

Each filter needs its own flat, shot at the correct focus for that filter. This is non-negotiable. An Ha flat won't work for OIII—the donuts are different sizes and positions.

Exposure times:
Narrowband flats require longer exposures than broadband flats because your light source (sky, panel) is much dimmer through narrow filters. What required 1 second for an L-band flat might need 5-10 seconds for Ha.

Target exposure level:
Aim for 40-50% histogram peak, same as broadband. If your light source is too dim, consider a brighter source or removing the flat filter briefly to expose through clear, then replacing the narrowband filter for actual flats.

9.4 Per-Filter Processing Workflow

Each narrowband filter is processed completely independently from sensor to integrated master.

Ha Processing Workflow

Characteristics:
Typically strongest signal, shortest exposures possible (10-15 minutes), most frames collected in a session.

Process:

1. Calibrate Ha lights with Ha flat (+ bias/dark)
2. SubframeSelector: Measure quality
3. StarAlignment: Register to best Ha frame
4. ImageIntegration: Stack with quality weighting

Common Ha issues:

Overexposure:
Ha is bright in emission nebulae. Bright core regions (like Orion's Trapezium) can saturate. Check histograms—if bright areas clip, you need shorter exposures for those regions (HDR workflow).

Sensitivity variations:
Different camera sensors have vastly different Ha sensitivity. Check your camera's QE curve.

OIII Processing Workflow

Characteristics:
Variable signal (strong in some targets, weak in others), often requires longer exposures than Ha (15-20 minutes).

Process:
Identical to Ha—completely independent preprocessing chain.

Common OIII issues:

Weak signal:
OIII is often the limiting factor in SHO imaging. If your target is OIII-faint, you need substantial integration time (10+ hours) to match Ha and SII depth.

Star color confusion:
OIII captures blue-green light. When combining for false color, OIII becomes the "blue" channel, but the stars it captured aren't naturally blue—this creates star color challenges in post-processing.

SII Processing Workflow

Characteristics:
Usually weakest signal, longest exposures needed (15-25 minutes), fewest frames per night.

Process:
Independent preprocessing, identical to Ha and OIII.

Common SII issues:

Very weak signal:
SII often requires 2-3× the integration time of Ha to achieve similar SNR. Many astrophotographers shoot 3 hours Ha, 4 hours OIII, 6 hours SII to balance the channels.

SNR challenges:
With weak signal and long exposures, noise becomes problematic. Aggressive integration (100+ frames if possible) is beneficial.

Diminishing returns:
At some point, additional SII integration yields minimal improvement. If your target is SII-weak (many are), accept that SII will be your limiting channel.

9.5 Preparing for Color Combination

After preprocessing, you have three independent masters: master_Ha.xisf, master_OIII.xisf, master_SII.xisf. These are still linear (unstretched) and ready for post-processing.

Matching Backgrounds Across Filters

Before color combination, backgrounds should be normalized. Different filters might have different background levels due to:

- Sky background variations (some sessions had more moon)
- Filter transmission differences
- Different total integration times

Process > BackgroundNeutralization can equalize backgrounds before color combination. Or handle this during initial stretching in post-processing.

Goal: When you stack RGB channels (or SHO mapped to RGB), backgrounds should match. Mismatched backgrounds create color casts.

Star Profile Matching

Different filters might have different FWHM due to:

- Chromatic focal shift (different wavelengths focus differently)
- Atmospheric dispersion (wavelength-dependent refraction)
- Different seeing conditions when filters were shot

Ideally:
Match FWHM across filters by applying slight blur to the sharper channels. This prevents one channel from dominating star sharpness in the color image.

Tools:
ConvolveMatrix or **MultiscaleLinearTransform** can subtly blur a channel to match others. This is typically done in post-processing, not preprocessing.

9.6 Common Narrowband Palettes

Understanding common color palettes helps you plan preprocessing. Each palette maps your three masters to RGB channels differently.

HOO (Bicolor)

Mapping:

- R: Ha
- G: OIII
- B: OIII

Characteristics:
Simplest narrowband workflow (only two filters needed). OIII does double duty in green and blue. Creates cyan/orange color schemes.

Preprocessing:
Only Ha and OIII needed. No SII required. Faster to collect full dataset.

SHO (Hubble Palette)

Mapping:

- R: SII
- G: Ha
- B: OIII

Characteristics:
Classic false-color palette popularized by Hubble images. Creates gold/blue aesthetics. Shows compositional information (different emission lines reveal different ionization zones).

Preprocessing:
All three filters required. SII's weakness often limits final image depth.

HOS (Alternative Palette)

Mapping:

- R: Ha
- G: OIII
- B: SII

Characteristics:
Different color aesthetic (pink/purple tones). Ha as red gives strong reds. SII in blue channel mutes blues (SII is weak).

Preprocessing:
Same as SHO—all three filters required.

Natural Palette

Mapping:

- R: Blend of SII and Ha (e.g., 0.75×SII + 0.25×Ha)
- G: OIII
- B: Blend of OIII and Ha (e.g., 0.9×OIII + 0.1×Ha)

Characteristics:
Attempts to represent emission line wavelengths more accurately. More complex but arguably more "realistic" false color.

Preprocessing:
All three filters with careful attention to relative SNR.

9.7 Narrowband Troubleshooting

"SII is too weak to be useful"

Cause:
Target is genuinely SII-faint, or insufficient integration time.

Solutions:

- Integrate SII dramatically longer (2-3× Ha time)
- Accept SII limitation and use HOO palette instead
- Use ImageIntegration with very gentle rejection (preserve every photon)
- Consider stacking darks too aggressively—easing dark subtraction can preserve faint signal

"Filters don't align perfectly"

Cause:
Chromatic focal shift means different wavelengths focus at slightly different positions.

Solution:

- Use StarAlignment to register OIII and SII masters to Ha master
- Small misalignments (0.5-1 pixel) are normal and acceptable
- Larger misalignments indicate focus problems between filter changes

"Ha overwhelming other channels"

Cause:
Ha naturally has 3-5× more signal than OIII or SII in many targets.

Solution:

- This is expected and normal
- During post-processing, scale channels appropriately (Ha might be reduced to 40% of its captured intensity to balance with OIII/SII)
- Consider shooting less Ha and more OIII/SII to balance integration times

"Background not neutral across filters"

Cause:
Different sky conditions, moon phases, or total integration times between filters.

Solution:

- Use BackgroundNeutralization before color combination
- Or normalize during initial stretching in post-processing
- Ensure calibration was successful (improper flats create background non-uniformity)

Key Takeaways from Chapter 9:

- Narrowband requires dramatically longer exposures than broadband (10-20+ minutes typical)
- Each filter (Ha, OIII, SII) preprocesses completely independently
- Monochrome cameras have substantial advantages for narrowband imaging
- Calibration is critical—dust shadows and amp glow are highly visible
- Each filter needs its own flat field (dust donuts differ by wavelength)
- SII is typically weakest and requires longest integration
- Different palettes (SHO, HOO, HOS) map filters to RGB differently
- Preprocessing goal: three high-quality linear masters ready for color combination
- Background matching and star profile consistency matter for clean color combination

Chapter 10 covers performance optimization and workflow efficiency—making PixInsight work faster and building repeatable, streamlined preprocessing workflows.

CHAPTER 10: PERFORMANCE OPTIMIZATION AND WORKFLOW EFFICIENCY

Making PixInsight Fast, Stable, and Repeatable

You've mastered the technical aspects of preprocessing. You understand calibration, registration, integration, and the intricacies of narrowband workflows. But there's a frustrating gap between knowing what to do and doing it efficiently.

Processing 150 frames shouldn't take three hours and multiple crashes. Running out of RAM halfway through integration shouldn't force you to restart. Repeating the same parameter settings for the fifth target this month shouldn't require consulting your notes every time. And when PixInsight hangs mysteriously at 87% progress, you shouldn't be left wondering whether to wait or force-quit.

This chapter transforms your preprocessing from a tedious, error-prone ordeal into a streamlined, reliable workflow. You'll learn how to configure PixInsight for maximum performance, avoid memory exhaustion and crashes, build reusable process icons and scripts, and troubleshoot the performance issues that plague even experienced users.

10.1 Understanding PixInsight's Resource Requirements

Memory: The Primary Bottleneck

PixInsight is memory-hungry. A single 26-megapixel image at 32-bit floating point requires approximately 100MB of RAM. But processing requires substantially more—PixInsight creates temporary copies, buffers, and intermediate results.

Memory multiplier rule: Processing requires roughly **10× the raw file size** in available RAM for comfortable operation.

Example calculation:

- Camera: 26MP (6248×4176 pixels)
- Raw file size: ~26MB (FITS format)
- Single image in memory (32-bit float): ~100MB (26MP × 4 bytes/pixel)
- Processing buffer needs: **~1GB per image**

When integrating 100 images, PixInsight doesn't load all 100 simultaneously, but it does need substantial RAM for:

- Current image being processed
- Reference image (for registration)
- Temporary buffers (rejection algorithms, normalization)
- Pixel statistics and weighting calculations

Recommended RAM by camera resolution:

Camera Resolution	Minimum RAM	Comfortable RAM	Ideal RAM
12MP (APS-C)	8GB	16GB	32GB
24MP (Full Frame)	16GB	32GB	64GB
45MP+ (High Res)	32GB	64GB	128GB
100MP+ (Medium Format)	64GB	128GB	256GB

CPU: Multi-Threading Benefits

PixInsight leverages multiple CPU cores effectively for most processes. More cores = faster processing, but with diminishing returns beyond 8-12 cores for typical workflows.

Best CPU characteristics:

- High single-thread performance (for single-image operations)
- 8-12 cores (sweet spot for multi-threaded batch operations)
- Modern architecture (newer CPUs have better instructions for floating-point math)

PixInsight thread configuration: Edit → Global Preferences → Process → Maximum Thread Count

Recommended setting: Number of physical CPU cores (not logical/hyperthreaded cores)

- 6-core CPU: Set to 6
- 8-core CPU: Set to 8
- 12-core CPU: Set to 12

Setting higher than physical core count can actually slow processing due to context-switching overhead.

Storage: Speed and Scratch Space

Working directory speed matters:

- **SSD (NVMe):** Excellent (5000+ MB/s read/write)
- **SSD (SATA):** Good (500 MB/s)
- **HDD:** Slow (100-150 MB/s)

Store working files on SSD. PixInsight frequently reads/writes temporary files during processing.

Scratch directory configuration: Edit → Global Preferences → Process → Scratch Directory

Set this to a fast SSD with substantial free space (100GB+ recommended for large projects).

10.2 File Format Optimization

XISF vs. FITS: Performance Comparison

FITS (Flexible Image Transport System):

- Universal standard (works everywhere)
- 32-bit floating point support
- Slower to read/write (older format, less optimization)
- Larger file sizes (no compression)

XISF (Extensible Image Serialization Format):

- PixInsight native format
- Faster read/write (modern design, optimized)
- Supports lossless compression (LZ4, Zlib)
- 64-bit floating point support (for extreme dynamic range)
- Smaller files (20-50% reduction with compression)

Performance difference: Loading 100 FITS files: ~45 seconds Loading 100 XISF files (compressed): ~25 seconds

Recommendation: Convert RAW files to XISF immediately after import for all subsequent processing.

Conversion process:

Process > Format Conversion > BatchFormatConversion

Input: RAW or FITS files

Output format: XISF

Compression: LZ4 (fast) or Zlib (smaller, slower)

Output directory: /working/XISF/

Work in XISF throughout preprocessing. Convert final integrated master to FITS if needed for external software compatibility.

Subframe File Management

Directory structure for efficiency:

/Project_M31/

/RAW/ (original camera files, archived)

/XISF/ (converted working files)

/Masters/ (bias, dark, flat masters)

/Calibrated/ (calibrated light frames)

/Registered/ (registered frames)

/Integrated/ (final masters)

/Drizzle/ (drizzle data files if applicable)

Why this helps:

- Clear separation prevents accidental overwriting
- Easy to re-run specific stages without starting over
- Intermediate results preserved for troubleshooting
- Organized for archival (keep RAW and Integrated, delete intermediate)

10.3 Avoiding Memory Exhaustion

Symptoms of Memory Problems

PixInsight running out of RAM manifests as:

- Processes slow to a crawl (excessive swapping to disk)
- Progress bar stuck at 90%+ for extended periods
- Crash with "Out of memory" error
- Windows becomes unresponsive
- Forced to terminate PixInsight

Prevention Strategies

1. Close Other Applications

Before heavy processing (ImageIntegration, StarAlignment batches), close:

- Web browsers (Chrome/Firefox can use 4-8GB)
- Photo editing software (Photoshop, Lightroom)
- Video players, games, unnecessary background apps

2. Process in Smaller Batches

Instead of integrating 200 frames at once:

- Integrate in groups of 50-70 frames
- Combine the resulting masters in a final integration

Batch integration workflow:

Batch 1: Frames 1-50 → master_batch1.xisf

Batch 2: Frames 51-100 → master_batch2.xisf

Batch 3: Frames 101-150 → master_batch3.xisf

Final: Integrate 3 masters → master_final.xisf

This reduces peak memory usage significantly.

3. Enable Scratch Files

Edit → Global Preferences → Process → Enable scratch files: ✓

Allows PixInsight to offload data to disk when RAM fills. Slower than pure RAM, but prevents crashes.

4. Reduce Preview Windows

Each open preview window consumes RAM. Close preview windows when not actively using them.

5. Clear Process Console History

The console accumulates log data. After heavy processing: Process → Clear Console

This frees memory used by logs.

Windows 24H2 Specific Issues

Recent Windows 11 24H2 update introduced memory management changes that cause PixInsight crashes for some users.

Symptoms:

- Random crashes during ImageIntegration
- "Access violation" errors
- Crashes at specific process stages

Workarounds:

- Roll back to Windows 11 23H2 (if possible)
- Disable Windows memory compression: Disable-MMAgent -MemoryCompression in PowerShell (admin)
- Increase PixInsight scratch file usage

- Process smaller batches
- Monitor PixInsight forum for official fixes

10.4 Building Reusable Process Icons

What Are Process Icons?

Process icons save configured processes (with all parameters) as reusable shortcuts. Instead of re-entering 15 parameters every time you run ImageCalibration, you drag a process icon onto your images.

Massive time saver for repeated workflows.

Creating Process Icons

Example: ImageCalibration for your camera

1. Configure ImageCalibration process completely:
   - Enable master bias, dark, flat
   - Set pedestal (20)
   - Set CFA pattern (RGGB)
   - Configure output directory
2. **Click the triangle icon** (bottom-left of process window)
3. Select **New Process Icon**
4. The process icon appears in the Process Explorer panel
5. **Rename it:** "ImageCal_Canon_R6_Bias_Dark_Flat"
6. **Drag icon onto images** to apply calibration instantly

Process Icon Library for Complete Workflow

Build a library of process icons for your typical workflow:

Calibration:

- ImageCal_Camera_Bias_Dark_Flat (full calibration)
- ImageCal_Camera_Dark_Flat (bias integrated with darks)
- ImageCal_Camera_Flat_Only (for when you lack darks)

Registration:

- StarAlign_Trans_Rot (standard registration)
- StarAlign_Trans_Rot_Drizzle (with drizzle data generation)
- StarAlign_Wide_Field (with distortion correction)

Integration:

- ImageInt_Standard (Average, Additive, Winsorized Sigma 4.0/2.5)
- ImageInt_Small_Set (Percentile clipping for <20 frames)
- ImageInt_Large_Set (Linear Fit for 100+ frames)

SubframeSelector:

- SubframeSel_Refractor (parameters for your refractor's image scale)
- SubframeSel_SCT (parameters for your SCT's image scale)

Backing Up Process Icons

Process icons are stored in PixInsight's settings directory. Back them up regularly:

Windows: C:\Users\[username]\AppData\Local\PixInsight\icons\
macOS: ~/Library/Application Support/PixInsight/icons/
Linux: ~/.local/share/PixInsight/icons/

Copy this entire folder to cloud storage or external drive. When reinstalling PixInsight or moving to a new computer, restore this folder to have all your process icons immediately available.

10.5 WBPP Advanced Optimization

WBPP Configuration for Speed

Multi-threaded execution: WBPP already leverages multi-threading, but you can optimize:

Number of parallel processes: Edit → Global Preferences → Process → Maximum parallel processes: **4-8**

Allows WBPP to calibrate multiple frames simultaneously (if RAM allows).

Memory-conscious settings: If running low on RAM during WBPP:

- Disable drizzle data generation (uses substantial RAM)
- Reduce "maximum stars" in registration settings (fewer stars = less memory)
- Process in smaller groups (manually divide lights into batches)

WBPP Resume After Crash

WBPP supports resuming if it crashes mid-process:

Before running WBPP:

- Enable "Save process log" in WBPP settings
- WBPP creates checkpoint files in output directory

If WBPP crashes:

1. Reopen WBPP
2. Reload your configuration
3. WBPP detects existing output files
4. Asks: "Resume from last checkpoint?"
5. Select **Yes** to continue where it left off

This prevents losing hours of progress if a crash occurs during Stage 4 of 5.

10.6 Workflow Scripts and Automation

PixInsight Scripting Basics

PixInsight supports JavaScript scripting for complete automation. You can script entire workflows from calibration through integration.

Script → Feature Scripts shows available scripts:

- BatchPreprocessing (predecessor to WBPP)
- SubframeSelectorMeasurements (automated quality assessment)
- FastRotation (quick image rotation)

Custom Workflow Script Example

Advanced users can write custom scripts. Example pseudocode for automated preprocessing:

```
// Load all light frames

var lights = LoadFiles("/data/lights/");

// Apply calibration using masters

for (var i = 0; i < lights.length; i++) {

 ApplyCalibration(lights[i], masterBias, masterDark, masterFlat);

}

// Measure quality with SubframeSelector

var qualities = MeasureQuality(lights);



// Reject poor frames (FWHM > 4.0)

var approved = FilterByFWHM(qualities, 4.0);

// Register to best frame

var reference = SelectBestFrame(approved);

RegisterFrames(approved, reference);

// Integrate with weighting

IntegrateFrames(approved, "Average", "Additive", "Winsorized");
```

Learning scripting:

- Study scripts in Script → Open Script Editor → Examples
- Read PixInsight scripting documentation (forum and wiki)
- Start simple (automate file renaming, batch conversions)
- Build complexity gradually

Third-Party Workflow Tools

Community-developed tools:

1. PreprocessLights Script (Darkblue) Simplified alternative to WBPP with clearer error messages and easier troubleshooting.

2. SpectrophotometricColorCalibration (SPCC) Automated color calibration using spectrophotometric data. Not preprocessing per se, but can be scripted into workflow.

3. BlinkTool Enhancements Scripts to automate satellite/airplane detection during manual review.

4. Batch File Converters Convert RAW → XISF with single script.

Search PixInsight forum for "workflow scripts" to discover community contributions.

10.7 Monitoring and Troubleshooting Performance

Process Console Analysis

The Process Console (bottom of screen) logs everything PixInsight does. Reading console output reveals:

Time benchmarks:

ImageIntegration: 150 files processed in 8m 23s

Average per file: 3.35s

Compare processing times between sessions. If suddenly slower, investigate:

- Disk speed (SSD vs. HDD)
- RAM usage (swapping to disk?)
- Background processes (antivirus scanning?)

Memory usage indicators:

Allocated memory: 24.3 GB

Peak usage: 28.7 GB (87% of physical)

If peak usage approaches 95%+ of physical RAM, you're hitting memory limits.

System Resource Monitoring

Windows: Task Manager (Ctrl+Shift+Esc) → Performance tab **macOS:** Activity Monitor **Linux:** htop or system monitor

Watch during processing:

- **CPU usage:** Should be 90-100% during multi-threaded operations
- **RAM usage:** Should stay below 90% (above that, swapping begins)
- **Disk usage:** SSD should show high activity during file I/O stages

Red flags:

- CPU at 100% but processing crawling: Thermal throttling (CPU overheating)
- RAM at 100%, disk activity maxed: Memory exhaustion, swapping to disk
- CPU low (<50%) during multi-threaded process: Thread count misconfigured

Common Performance Issues

"ImageIntegration stuck at 90%":

- Actually still processing (rejection algorithms intensive at final stages)
- Check RAM usage—if maxed, it's swapping to disk (wait it out)
- If waited 30+ minutes with no progress, likely crashed (force quit and restart)

"StarAlignment taking forever":

- Too many stars detected (hundreds of thousands)
- Reduce maximum stars in settings (5000 is plenty)
- Or increase detection scales (fewer small stars detected)

"WBPP hangs during registration":

- Detection scales wrong (measure star sizes manually, override auto-detection)
- One frame radically different (clouds, focus failure) causing registration failure
- Check console for specific error messages

"Crashes during integration":

- Memory exhaustion (reduce batch size)
- Windows 24H2 issue (see workarounds above)
- Corrupted frame causing crash (identify via Blink, exclude from integration)

Key Takeaways from Chapter 10:

- Memory is the primary bottleneck—10× raw file size in RAM recommended
- Convert RAW to XISF immediately for 40%+ speed improvement
- Process in smaller batches to avoid memory exhaustion
- Build library of process icons for repeated operations
- Enable scratch files to prevent out-of-memory crashes
- Configure thread count to match physical CPU cores
- Use structured directory organization for efficient file management
- WBPP can resume after crashes if configured properly
- Monitor system resources during processing to identify bottlenecks
- Windows 24H2 has known issues—workarounds available
- Scripting enables complete workflow automation for advanced users

Conclusion: From Overwhelmed to Optimized

You've progressed from understanding basic preprocessing concepts through mastering complex multi-session workflows, narrowband imaging, and performance optimization. PixInsight no longer feels like a black box—you understand each stage, why it matters, and how to troubleshoot when things go wrong.

The preprocessing workflows in this book form the foundation of exceptional astrophotography. Perfect preprocessing doesn't guarantee a stunning final image, but poor preprocessing guarantees disappointment no matter how skilled your post-processing.

Now, turn your calibrated, registered, and integrated masters into the deep-sky masterpieces you've envisioned. The stars are waiting.

Clear skies and happy stacking!

APPENDICES

APPENDIX A: Quick Reference Cards

A.1 Master Calibration Frame Quick Reference

Frame Type	How Many	Settings	Key Parameters
Bias	100 (50-200)	Shortest exposure, same ISO/gain as lights	Average combination, Percentile clipping
Dark	50 (30-100)	Match exposure AND temperature ±2°C	Average, Sigma clipping 4.0/3.0
Flat	30-50 (20 min)	Histogram peak 40-50%, match filter/focus	Average, **Multiplicative normalization**

Critical Reminders:

- Temperature tolerance: ±2°C excellent, ±5°C acceptable, ±10°C problematic
- Flats MUST be calibrated with bias/dark before stacking
- One flat per filter (chromatic focal shift makes this non-negotiable)

A.2 ImageIntegration Settings by Scenario

Scenario	Frame Count	Combination	Normalization	Weights	Rejection	Sigma Low/High
Standard	30-100	Average	Additive	SSWEIGHT	Winsorized Sigma	4.0 / 2.5
Small Set	10-20	Average	Additive	Noise Eval	Percentile	0.2 / 0.1
Large Set	100+	Average	Additive	SSWEIGHT	Linear Fit	3.0
Satellite-Heavy	Any	Average	Additive	Noise Eval	Winsorized Sigma	4.0 / 2.0
Multi-Night	Masters only	Average	Adaptive	Noise Eval	None	N/A

A.3 StarAlignment Registration Models

Model	Use When	Speed	Complexity
Translation	No rotation, simple tracking	Fastest	Lowest
Translation + Rotation	Standard deep-sky (recommended)	Fast	Low
Translation + Rotation + Scale	Different instruments, atmospheric refraction	Medium	Medium
Full Projective + Distortion	Wide-field with barrel/pincushion distortion	Slow	High

Detection Scales Rule: Measure star diameter in reference frame, set scales to bracket (e.g., 3px stars → use 2-6 or 3-7)

A.4 Exposure Time and SNR Improvement

Frame Count	SNR Improvement	Equivalent Single Exposure
4 frames	2×	4× exposure time
9 frames	3×	9× exposure time
16 frames	4×	16× exposure time
25 frames	5×	25× exposure time
100 frames	10×	100× exposure time

Formula: SNR improvement = √N (where N = number of frames)

A.5 Memory Requirements by Camera Resolution

Camera Resolution	Raw File Size	Processing Memory Needed	Recommended System RAM
12MP (APS-C)	~12MB	~120MB per image	16GB minimum
24MP (Full Frame)	~26MB	~260MB per image	32GB recommended
45MP (High Res)	~45MB	~450MB per image	64GB recommended
61MP	~61MB	~610MB per image	64GB minimum
100MP+ (Medium Format)	~100MB +	~1GB+ per image	128GB+

Rule of thumb: Processing needs ~10× raw file size in available RAM

A.6 Narrowband Filter Characteristics

Filter	Wavelength	Typical Bandpass	Relative Signal Strength	Typical Exposure
Ha (Hydrogen-Alpha)	656nm	3nm, 6nm, 7nm	Strongest (1.0×)	10-15 min
OIII (Oxygen-III)	496/501nm	3nm, 6nm	Moderate (0.5-0.8×)	15-20 min
SII (Sulfur-II)	672nm	3nm, 6nm	Weakest (0.3-0.5×)	15-25 min

Color Palette Mappings:

Palette	Red	Green	Blue
SHO (Hubble)	SII	Ha	OIII
HOO (Bicolor)	Ha	OIII	OIII
HOS	Ha	OIII	SII

A.7 Common FITS Header Keywords

Keyword	Meaning	Example Value
IMAGETYP	Frame type	LIGHT, DARK, FLAT, BIAS
EXPTIME	Exposure time (seconds)	300.0
GAIN	Camera gain setting	139
ISOSPEED	ISO setting (DSLR)	800
SET-TEMP	Target temperature (°C)	-10.0
CCD-TEMP	Actual temperature (°C)	-10.2
FILTER	Filter name	Ha, OIII, Luminance
XBINNING / YBINNING	Pixel binning	1 (no binning)

APPENDIX B: Troubleshooting Flowcharts

B.1 Calibration Troubleshooting Flowchart

Calibrated frames look wrong

|

├─→ Too dark / negative pixels

| |

| ├─→ Check dark exposure/temperature match

| ├─→ Verify EXPTIME and SET-TEMP in FITS headers

| ├─→ Disable dark optimization

| └─→ Reshoot darks if mismatch confirmed

|

├─→ Strange patterns / artifacts

| |

| ├─→ Examine master flat for gradients

| ├─→ Test calibration without flat

| ├─→ Check FITS orientation setting (must match)

| └─→ Reshoot flats with better illumination

|

├─→ Wrong colors (OSC only)

| |

| ├─→ Verify CFA pattern setting (RGGB, BGGR, etc.)

| ├─→ Check flat light source (must be white)

| └─→ Ensure masters were Bayer CFA format

|

└─→ Calibration worse than raw

|

├─→ Check master quality (noisy masters add noise)

├─→ Test selective calibration (bias+flat only)

└─→ Sometimes no calibration is better than bad calibration

B.2 Registration Troubleshooting Flowchart

Registration fails or looks wrong

|

├─→ "No stars detected"

| |

| ├─→ Detection scales wrong (most common)

| ├─→ Measure actual star FWHM in reference

| ├─→ Set scales to bracket measured size

| └─→ Click eyeball icon to preview detections

|

├─→ "Failed to register X images"

| |

| ├─→ Frames too different (clouds, tracking errors)

| ├─→ Inspect failed frames in Blink

| ├─→ Reject if genuinely bad

| └─→ Try wider detection scales

|

├─→ Images register but look misaligned

| |

| ├─→ Over-fitting (model too complex)

| ├─→ Reduce model complexity (Translation + Rotation only)

| ├─→ Increase detection scales

| └─→ Verify reference frame is sharp and artifact-free

|

└─→ Misalignment at edges

|

├─→ Optical distortion not modeled

├─→ Enable distortion correction

└─→ Try Full Projective model

B.3 Integration Troubleshooting Flowchart

Integration results unsatisfactory

|

├─→ Stars not sharp / bloated

| |

| ├─→ Registration misalignment (check registered frames)

| ├─→ Poor seeing frames included (check FWHM distribution)

| └─→ Reference frame selection poor (choose sharpest)

|

├─→ Artifacts / streaks visible

| |

| ├─→ Rejection too gentle (satellite/cosmic ray not removed)

| ├─→ Increase rejection aggressiveness (lower sigma high)

| └─→ Check rejection maps (should show scattered hits)

|

├─→ Background uneven / patchy

| |

| ├─→ Normalization method wrong (try Adaptive)

| ├─→ Frames have very different backgrounds

| └─→ Consider LocalNormalization before integration

|

├─→ Result darker/dimmer than expected

| |

| ├─→ Over-rejection (too much signal rejected)

| ├─→ Ease rejection parameters (higher sigma values)

| └─→ Check frame count (some frames rejected entirely?)

|

└─→ Noise higher than expected

|

├──→ Too few frames (need more integration time)

├──→ Weights not applied (check SSWEIGHT exists)

└──→ Poor quality frames dragging down result

B.4 WBPP Troubleshooting Flowchart

WBPP fails or produces poor results

|

├──→ "All frames rejected"

| |

| ├──→ Check calibrated frames manually

| ├──→ If calibration wrong → master calibration issue

| ├──→ If calibration fine → disable subframe weighting

| └──→ Verify FITS headers match (EXPTIME, FILTER, GAIN)

|

├──→ "Registration failed for most frames"

| |

| ├──→ Detection scales wrong (measure manually, override)

| ├──→ Check WBPP console for specific errors

| └──→ Test registration manually on sample frames

|

├──→ Calibrated frames terrible

| |

| ├──→ Master calibration wrong (temp/exposure mismatch)

| ├──→ Check masters in /output/masters/

| ├──→ Verify FITS headers match between masters and lights

| └──→ Test calibration manually with ImageCalibration

|

├──→ Process hangs / no progress

| |

| ├──→ Memory exhaustion (check Task Manager)

| ├──→ Convert RAW to XISF first (reduces memory)

| ├──→ Process smaller batches

| └──→ Close other applications

|

└──→ Strange output / unexpected files

|

├──→ Review WBPP console log carefully

├──→ Check output directory structure

└──→ Compare against expected workflow stages

APPENDIX C: Camera-Specific Workflows

C.1 DSLR / Mirrorless Workflow

Key Characteristics:

- Bayer CFA sensor (must debayer after calibration)
- ISO settings instead of gain
- Temperature-dependent darks (varies with ambient)
- Built-in RAW format

Preprocessing Steps:

1. **Import RAW files**
   - Format Explorer: Set RAW format to "Raw Bayer CFA"
   - Convert to XISF immediately for speed
2. **Create master calibration frames**
   - Bias: 100 frames, shortest exposure, same ISO
   - Darks: 50 frames, match exposure AND temperature (±5°C acceptable for DSLRs)
   - Flats: 30-50 frames, calibrated with bias/dark
3. **Calibrate lights**
   - ImageCalibration: Enable bias, dark, flat
   - **Set CFA pattern** (usually RGGB, verify for your camera)
   - Pedestal: 20
   - Output still Bayer CFA (not debayered yet)
4. **Debayer calibrated frames**
   - Process > Debayer > Debayer
   - Algorithm: VNG or AHD
   - Apply to all calibrated frames
5. **Quality assessment**
   - SubframeSelector on debayered frames
   - Set camera parameters (pixel size, focal length, gain)
6. **Register**
   - StarAlignment: Translation + Rotation
   - Detection scales based on star size
7. **Integrate**
   - ImageIntegration: Average, Additive, SSWEIGHT

Camera-Specific Settings:

Camera Model	Sensor Size	Pixel Size	Typical CFA Pattern
Canon EOS R6 II	Full Frame	6.2μm	RGGB
Canon EOS R7	APS-C	3.20μm	RGGB
Nikon Z6 II	Full Frame	5.94μm	RGGB
Sony A7 IV	Full Frame	4.28μm	RGGB
Nikon Z50	APS-C	3.88μm	RGGB

C.2 OSC Dedicated Astro Camera Workflow

Key Characteristics:

- Bayer CFA sensor (must debayer)
- Gain settings (electrons/ADU)
- TEC cooling (stable temperature)
- FITS output format

Preprocessing Steps:

1. **Organize FITS files**
   - Already in FITS, convert to XISF for speed
   - FITS headers include temperature, gain, exposure
2. **Create master calibration frames**
   - Bias: 100 frames
   - Darks: 50 frames, **exact temperature match** (±2°C)
   - Flats: 30-50 frames per filter
3. **Calibrate lights**
   - ImageCalibration: Same as DSLR
   - Set CFA pattern (verify in camera manual)
   - Temperature matching critical
4. **Debayer → Register → Integrate**
   - Same as DSLR workflow

Popular OSC Astro Cameras:

Camera Model	Sensor Size	Pixel Size	Gain (e-/ADU)	Cooling

ZWO ASI533MC Pro	APS-C	3.76μm	1.0 @ Gain 100	Yes (-35°C)
ZWO ASI2600MC Pro	Full Frame	3.76μm	0.5 @ Gain 100	Yes (-35°C)
QHY268C	APS-C	3.76μm	Variable	Yes
Player One Apollo-M MAX	Micro 4/3	3.76μm	0.5 @ Gain 100	Yes (-35°C)

C.3 Cooled Monochrome Workflow

Key Characteristics:

- No CFA (no debayering needed)
- Per-filter imaging (LRGB or narrowband)
- TEC cooling (precise temperature control)
- Highest sensitivity

Preprocessing Steps:

Per Filter (L, R, G, B, Ha, OIII, SII):

1. **Organize by filter**
   - Separate directories: /Lights_L/, /Lights_R/, etc.
   - Each filter processed completely independently
2. **Create master calibration frames**
   - Bias: Shared across all filters (100 frames)
   - Darks: Shared if exposure times match (50 frames per exposure time)
   - **Flats: One per filter** (critical—chromatic focal shift)
3. **Calibrate lights** (per filter)
   - ImageCalibration with appropriate flat for this filter
   - No CFA pattern needed (monochrome)
   - No debayering step
4. **SubframeSelector** (per filter)
   - Quality assessment on this filter's frames only
5. **Register** (per filter)
   - StarAlignment to best frame for this filter
6. **Integrate** (per filter)
   - Result: master_light_L.xisf, master_light_R.xisf, etc.

After preprocessing all filters:

- Combine in post-processing (LRGB combination, narrowband palettes)

Popular Monochrome Cameras:

Camera Model	Sensor Size	Pixel Size	Gain	Cooling	Typical Use
ZWO ASI533 MM Pro	APS-C	3.76 μm	1.0 @ G100	-35°C	LRGB, Narrow band
ZWO ASI1600 MM Pro	Micro 4/3	3.8μm	0.18 @ G139	-35°C	Fast narrow band
QHY600M	Full Frame	3.76 μm	Variable	Yes	High-res LRGB
ZWO ASI6200 MM Pro	Full Frame	3.76 μm	Variable	-35°C	High-res narrow band

APPENDIX D: Process Parameter Tables

D.1 ImageCalibration Parameters

Parameter	Options	Recommended	Notes
Enable master bias	On/Off	On	Unless integrated with darks
Enable master dark	On/Off	On	Always use darks
Enable master flat	On/Off	On	Critical for dust/vignetting
Optimize darks	On/Off	**Off**	Only if exposure mismatch unavoidable
Pedestal	0-100	10-20	Prevents negative pixels
Output directory	Path	/Calibrated/	Organized structure
CFA pattern	RGGB,	Camera-specific	Verify in manual

	BGGR, etc.		(OSC/DSLR only)
Output file format	FITS, XISF	XISF	Faster processing

D.2 SubframeSelector Parameters

Parameter	Options	Recommended	Notes
Subframe scale	arc-sec/pixel	Calculate: (pixel_size / focal_length) × 206.265	Critical for FWHM measurement
Camera gain	e-/ADU	Camera-specific	Check camera specs
Structure layers	1-7	5	Star detection sensitivity
Noise layers	0-5	0-2	Lower = more sensitive
Sensitivity	0.0-1.0	0.1	Lower detects more stars
Weighting preset	Default, Galaxy, Planetary Nebula, Star Field	Default	Use Galaxy for extended objects

D.3 StarAlignment Parameters

Parameter	Options	Recommended	Notes
Registration model	See A.3	Translation + Rotation	Standard deep-sky
Detection scales	pixels	Bracket measured FWHM	Most common error source
Sensitivity	0.0-1.0	0.1	Lower = more stars
Peak response	0.0-1.0	0.8	Star detection threshold
Maximum stars	Number	2000-5000	More = slower but more accurate

Distortion correction	On/Off + Order	Off / Order 2 if enabled	Wide-field only
Generate drizzle data	On/Off	On if dithered	Required for DrizzleIntegration
Drop shrink	0.5-1.0	0.9	Lower = more aggressive sharpening

D.4 ImageIntegration Parameters

Parameter	Options	Recommended	Notes
Combination	Average, Median, Min, Max	Average	Optimal SNR
Normalization	No, Additive, Multiplicative, Adaptive	Additive	Standard deep-sky
Weights	None, FITS keyword, Noise evaluation, etc.	FITS keyword (SSWEIGHT)	From SubframeSelector
Rejection algorithm	See A.2	Winsorized Sigma Clipping	Most robust
Sigma low / high	Values	4.0 / 2.5	Standard settings
Range low / high	0.0-1.0	0.0 / 1.0	Don't clip valid data
Generate integrated image	On/Off	On	Main output
Generate drizzle data	On/Off	On if dithered	For super-resolution

APPENDIX E: Community Resources

E.1 Official PixInsight Resources

Official Website: https://pixinsight.com

Official Documentation: https://pixinsight.com/doc/

Official Forum: https://pixinsight.com/forum/

- Most comprehensive troubleshooting resource
- Active developer participation
- Search before posting (answers often exist)

YouTube Channel: https://youtube.com/@OfficialPixInsight

- Official tutorials and demonstrations
- Process deep-dives

E.2 Recommended Learning Resources

Light Vortex Astronomy (Rogelio Bernal Andreo):

- Website: https://lightvortexastronomy.com
- Comprehensive PixInsight tutorials
- Deep technical explanations
- Advanced techniques

Harry's Astroshed:

- Website: https://www.harrysastroshed.com
- Beginner-friendly tutorials
- Focus on practical workflows

Cuiv, The Lazy Geek:

- YouTube channel with PixInsight tutorials
- Emphasis on automation and efficiency

PixInsight User Groups (Facebook):

- Active communities for questions and image critiques
- Real-world problem-solving

Cloudy Nights Astrophotography Forum: https://www.cloudynights.com/forum/15-astrophotography/

- Large community, PixInsight subforum
- Equipment and workflow discussions

E.3 Essential Third-Party Scripts

EZ Processing Suite:

- Simplifies common workflows
- Available on PixInsight forum

Starnet++ Script:

- Star removal for processing (post-preprocessing)
- https://www.starnetastro.com

BlurXTerminator:

- AI-powered deconvolution (post-preprocessing)
- https://www.rc-astro.com/resources/BlurXTerminator/

NoiseXTerminator:

- AI-powered noise reduction (post-preprocessing)
- https://www.rc-astro.com/resources/NoiseXTerminator/

E.4 Image Scale Calculators

Online calculators for determining arc-seconds/pixel:

Astronomy.tools: https://astronomy.tools/calculators/field_of_view/

- Comprehensive field of view calculator
- Includes resolution, sampling, etc.

Formula:

Image Scale (arc-sec/pixel) = (Pixel Size in μm / Focal Length in mm) × 206.265

Example:

- Pixel size: 3.76μm
- Focal length: 480mm
- Image scale = (3.76 / 480) × 206.265 = **1.62 arc-seconds/pixel**

E.5 Gain and Read Noise Resources

Camera gain and read noise measurements:

ZWO Cameras:

- Official specifications on product pages
- https://astronomy-imaging-camera.com

QHY Cameras:

- Specifications in user manuals
- https://www.qhyccd.com

Community-measured sensor data:

- Cloudy Nights camera reviews
- YouTube reviews with test data

APPENDIX F: Glossary of Terms

Additive Normalization: Integration normalization method that equalizes background levels by adding/subtracting offsets.

Amp Glow: Electronic glow visible in sensor corners during long exposures, caused by amplifier circuitry.

Bayer CFA (Color Filter Array): Mosaic of red, green, blue filters over sensor pixels in color cameras (typically RGGB pattern).

Bias Frame: Shortest possible exposure (covered sensor) capturing electronic offset and read noise pattern.

Calibration: Process of removing sensor-specific artifacts (bias, dark current, vignetting, dust) from light frames.

Cosmic Ray: High-energy particle hit creating bright pixel(s) in single frame, removed via rejection during integration.

Dark Current: Thermal electrons accumulating in pixels during exposure, temperature-dependent.

Dark Frame: Covered-sensor exposure matching light frame exposure time and temperature, capturing bias + dark current + hot pixels.

Debayering: Converting Bayer CFA data into RGB image by interpolating missing color information at each pixel.

Detection Scales: Range of star sizes (in pixels) that StarAlignment searches for during registration.

Drizzle: Super-resolution technique combining dithered frames to increase resolution beyond pixel size.

Eccentricity: Measure of star roundness (0 = perfect circle, 1 = line), indicates tracking quality.

Flat Frame: Image of uniformly illuminated field capturing vignetting, dust shadows, and pixel response variations.

FWHM (Full Width Half Maximum): Diameter of star profile at half its peak intensity, measured in pixels or arc-seconds. Lower = sharper.

Histogram: Graph showing distribution of pixel brightness values in image.

Hot Pixel: Defective pixel with excessive dark current, appears as bright dot even with sensor covered.

Integration: Stacking/combining multiple registered frames into single master, improving SNR.

Linear Data: Image data before stretching/curves, proportional to photons captured. All preprocessing done on linear data.

LocalNormalization: Spatially-varying background normalization handling gradients that differ between frames.

Master Frame: Stacked/integrated calibration frame (master bias, master dark, master flat).

Meridian Flip: 180° mount rotation when crossing meridian, requiring registration to handle rotation.

Multiplicative Normalization: Integration normalization method that scales entire frame intensity (required for flat field integration).

Narrowband: Filters passing extremely narrow spectral bands (3-7nm) centered on emission lines (Ha, OIII, SII).

Normalization: Equalizing background levels or overall intensity between frames before integration.

OSC (One-Shot Color): Camera with Bayer CFA capturing all colors simultaneously in single exposure.

Pedestal: Small positive offset added during calibration to prevent negative pixel values after dark subtraction.

QE (Quantum Efficiency): Percentage of photons hitting sensor that generate electrons (signal). Higher = more sensitive.

Read Noise: Electronic noise added during pixel value readout from sensor.

Reference Frame: Highest-quality frame used as alignment target during registration.

Rejection: Outlier removal during integration (satellites, cosmic rays, airplanes).

Registration: Aligning frames with sub-pixel precision so stars overlap exactly.

Sigma Clipping: Rejection algorithm removing pixel values more than N standard deviations from median.

SNR (Signal-to-Noise Ratio): Ratio of signal (real photons) to noise (random variations). Higher = cleaner image.

SSWEIGHT: Quality weight assigned by SubframeSelector, stored in FITS header, used during integration.

StarAlignment: PixInsight process for registering (aligning) frames by detecting and matching star patterns.

SubframeSelector: PixInsight process for measuring frame quality (FWHM, eccentricity, SNR) and calculating weights.

Vignetting: Darkening at image edges/corners due to optical design, removed by flat field calibration.

WBPP (WeightedBatchPreprocessing): Automated PixInsight script performing complete preprocessing workflow (calibration through integration).

XISF (Extensible Image Serialization Format): PixInsight native file format, faster and more efficient than FITS.

INDEX

W

X

www.ingramcontent.com/pod-product-compliance
Lightning Source LLC
Chambersburg PA
CBHW061222240926
51846CB00006B/1320
* 9 7 8 5 5 6 2 7 9 9 0 7 4 *